Frontiers in Ecoacoustics

Volume 1

Series Editor
Almo Farina, Department of Pure and Applied Sciences, University of Urbino, Urbino, Pesaro-Urbino, Italy

Ecoacoustics is a recent ecological discipline that aims to investigate the role of sounds in the ecological processes from individual animal species, to population and communities. It's competence is rapidly growing contributing to fill several gaps in the knowledge of natural phenomena. To day, ecoacoustics offers a set of robust theories to build efficient models to interpret complex ecological dynamics and to explain the adaptations of species to a changing world. Contemporarily, the availability of well tested metrics allows mining and processing a large amount of data collected by autonomous devices (digital recorders).

The Springer book series in ecoacoustics intends to publish theoretical, methodological and descriptive investigations, projects and solutions for a sustainable land cover management thanks to the contribution offered by experts from all the field of ecoacoustics. With the book series in ecoacoustics we intend to offer an agile editorial platform and permanent forum to scholars from all the world to illustrate and discuss new ideas and ecoacoustics theories, to summarize results on soundscapes from terrestrial, freshwater and marine biomes, and to make extensive reports on the ecoacoustics methodologies applied to environmental assessment and resources monitoring.

The aim of this book series is to increase the popularity of ecoacoustics as a discipline that plays the role of bridge between ecology, ecosemiosis, animal behaviour, bioacoustics and psychoacoustics.

The series is conceived as a flexible tool with the strategy to offer the best news in the ecoacoustics field also discussed in workshops, symposia and congresses.

The series will encourage interdisciplinary studies to better understanding the relationship between other sensory-based environments, like visual, thermal, olfactory, tactile-scapes.

The series will devote particular attention to principles, theories and methods in ecoacoustics. Contemporarily the series intend to promote a comparative analysis of soundscapes from different biogeographic regions and land management regimes.

The effect of noise on animal communities and people, land cover change and acoustic habits, traditional versus industrial farming and their effects on acoustic communities, the cultural and recreational value of soundscapes, are some of the topics that could find space into the series.

More information about this series at https://link.springer.com/bookseries/16630

Almo Farina • Peng Li

Methods in Ecoacoustics

The Acoustic Complexity Indices

Springer

Almo Farina
Department of Pure and Applied Sciences
University of Urbino
Urbino, Pesaro-Urbino, Italy

Peng Li
Division Sleep and Circadian Disorders
Brigham and Women's Hospital
Boston, MA, USA

ISSN 2730-6542 ISSN 2730-6550 (electronic)
Frontiers in Ecoacoustics
ISBN 978-3-030-82176-0 ISBN 978-3-030-82177-7 (eBook)
https://doi.org/10.1007/978-3-030-82177-7

This Springer imprint is published by the registered company Springer Nature Switzerland AG
The registered company address is: Gewerbestrasse 11, 6330 Cham, Switzerland

Foreword

Sounds, originated by vibrations of particles and cognitively transformed in a mental representation, concur to the structure and to the dynamics of natural and human-modified systems.

In particular, sounds are vehicles of information that animals either utilize actively with emission of specific acoustic signals or passively by listening the broadcasted acoustic signals of other individuals and species.

Recently, ecoacoustics has been recognized as an important ecological discipline in the exploration of the semiotic relationships between organisms and in the assessment of the environmental complexity and habitat quality, as well. For this, a broad range of competencies and uses are recognized in ecoacoustics. Moreover, the development of efficient metrics supported by automatic procedures allows the assessment of the sonic complexity, the evaluation of the level of intactness of the environment, or, anticipating, the effects of climatic change on species, habitats and landscapes.

In this book, after a short concise introduction to principles, theories and models in ecoacoustics, the Acoustic Complexity Indices (ACIs) that represent the most used ecoacoustic metrics in terrestrial and aquatic environments are described in detail.

A specific software (*SonoScape*) that processes ACIs and the other associated metrics developed in MATLAB® environment is offered as open source script, associated to detailed instructions for its use. Examples and exercises of application of ACI in the ecological survey of landscapes are also provided.

Urbino Italy Almo Farina

Boston MA USA Peng Li

Contents

Chapter 1
An Introduction to Ecoacoustics

Abstract Ecoacoustics is a new discipline that aims to investigate the ecological role of sounds of geological, biophonic, and anthropogenic origin. Its development has been favored by new robust theoretical principles associated to efficient metrics for data processing and by the availability of autonomous acoustic recorders to collect a great number of acoustic files at different temporal and geographical scale.

The double role of sound as a semiotic tool to communicate and as ecological proxy of environmental conditions to select habitats and to navigate represents the ideal condition for a rapid development of this discipline. The transformation of latent vibrations as generators of any typology of sound, a clear semiosis that recognizes a sonoscape as the component of the original vibroscape sensed by organisms, and a soundscape as the portion of sonoscape interpreted by individual species are three components of the sonic domain. Sonotope and soundtope, respectively, are sensed and interpreted patches with which soniferous species and acoustic communities interact in a spatial sonic mosaic.

The Morphological Adaptation Hypothesis, the Acoustic Adaptation Hypothesis, the Acoustic Niche Hypothesis, the Acoustic Community Hypothesis, and the Acoustic Habitat Hypothesis represent the theoretical fundaments of ecoacoustics.

An acoustic community is defined as the collection of soniferous species acoustically active in space and time. Such aggregation of soniferous species determines a sonic environment variable in space and time and represented by sonic matrices (on which to apply ecoacoustics metrics) after a process of migration from a temporal domain to a frequential domain via a Fourier Transform. A large portion of ecoacoustic investigations focuses on the role of noise (especially of anthropogenic origin) on behavioral and ecological processes in terrestrial and in aquatic ecosystems.

To describe the complex sonic domain where an originator vibroscape evolves into several distinct objects obtained after a latent, sensed, and interpreted semiosis requires the development of a dedicated narrative. Sonoscape is the result of a sensed vibroscape, and a soundscape is obtained from an interpreted sonoscape. Sonotopes represent the "geographical" elements composing a sonoscape, and their detection is obtained by the deployment of sound recorders according to a configuration that enhances the spatial heterogeneity. Sonotopes are the spatial unit of a sonic information system and represent the link with the geographical character of

A. Farina, P. Li, *Methods in Ecoacoustics*, Frontiers in Ecoacoustics 1,
https://doi.org/10.1007/978-3-030-82177-7_1

a landscape. Every sonotope is characterized by species-specific sonic and acoustic signatures. The former (species-specific sonic signature) is the result of a sensed vibroscape and the latter (species-specific acoustic signature) of the interpretation of sonic signals.

Ecoacoustic events obtained by coding three metrics ($ACIf_t$, $ACIf_{t\ evenness}$, and $ACIt_{f\ evenness}$) are an attempt to discretize acoustic signals into functional or statistical distinct units.

Keywords Ecoacoustics · Vibroscape · Sonoscape · Sonotope · Soundscape · Soundtope · Morphological adaptation hypothesis · Acoustic adaptation hypothesis · Acoustic niche hypothesis · Acoustic habitat hypothesis · Acoustic community hypothesis

1.1 Ecoacoustics: The Definition of a New Scientific Discipline

Ecoacoustics is "a theoretical and applied discipline that studies sound along a broad range of spatial and temporal scales in order to tackle biodiversity and other ecological questions" (Sueur & Farina, 2015). In the recent decades, new theoretical foundations associated to the advent of innovative technologies that allow the automatic collection of large set of data for long periods of time and an efficient data processing by using new algorithms (Towsey et al., 2014a, b; Farina et al., 2018, 2021a, b) handled by dedicated software (f.i., Seewave, Sueur et al., 2008; Sueur, 2018) have boosted the quick development of ecoacoustics and offered new tools to investigate the ecological complexity.

Ecoacoustics has been developed favored by:

1. Robust theoretical bases that have considered the ontological aspects of soundscape and their components (Pijanowski et al., 2011a, b; Farina, 2014; Farina et al., 2021a, b).
2. Availability of autonomous digital recorders that can be deployed for some time in the field (e.g., Agranat, 2007, 2009, http://www.iinsteco.org/soundscape_explorer/, https://www.openacousticdevices.info/audiomoth). The transaction from indoor recorders usable in field only with favorable atmospheric conditions and for a short period to autonomous recorders to deploy in the wild with resistance to climatic stresses represents the turning point in the development of ecoacoustics. Recently, the availability of autonomous recorders at a lower cost (Hill et al., 2019) represents a further opportunity to expand the ecoacoustic research.
3. Possibility to store sounds into permanent and large memory for further processing.
4. Availability of powerful algorithms to process data (Pieretti et al., 2011; Depraetere et al., 2012; Sueur et al., 2014; Ross et al., 2021).
5. New software that can organize and process more efficiently large set of data.

The ecoacoustic practice concerns the acquisition of sonic data from the environment. Such (raw) data are the base to apply different ecoacoustics models in an attempt to pair up sonic phenomena to biological and ecological processes.

Many concepts that are used in ecoacoustics can be verified experimentally or at least can be utilized to better explain biological and ecological processes: The Acoustic Complexity Indices (ACIs) have been formalized to reach this goal.

In turn, ACI metrics are powered by ***SonoScape***, presented for the first time in this book, with the aim to describe new ecoacoustic processes and validate functional models of the role of sound in ecological dynamics (Farina et al., 2021a, b).

1.2 Short History of Ecoacoustics

Before 2014, the studies on the environmental sounds were labeled as "soundscape ecology" (Pijanowski et al., 2011a, b), but after the discussion emerged during a first congress on the role of environmental sounds organized in Paris in that year by Jerome Sueur and one of us (AF), a remote polling between scholars attributed to "Ecoacoustics" the role of an umbrella discipline. Soundscape ecology was considered a subordinate domain with competencies restricted to the relationships between landscapes and soundscapes (Farina, 2018). Today, ecoacoustics appears as a robust discipline with a wide range of interests and applications from biodiversity monitoring (Sueur et al., 2014; Stowell & Sueur, 2020) to effect of sonic pollution (Naguib, 2013), and to the effect of climate change on livings (Krause & Farina, 2016).

1.3 Sound Definition and Categories

Sounds are obtained when vibrations from an elastic body pass through air (airborne) or water (water-borne) and are received by the animal-hearing organs. After a codification, such vibrations are transmitted to the central nervous system that returns a mental sensation called sound. Sound can be defined as the sensation caused by pressure variations in the air.

The noun sound can be associated to the adjective "sonic" or to the adjective "acoustic". Although the difference between the adjective sonic (Latin origin: sound = sonus) and acoustic (Greek: sound = ήχος [ichos]) seems really negligible, in this booklet, we adopt the adjective "sonic" when related to the entire set of physical vibrations sensed. The term acoustic is related to the part of vibrations interpreted by organism's hearing. In the same way, we adopt the term sonic signature to describe the characteristic of an environment in terms of sensed vibrations. We will use acoustic signature with the meaning of the part of a vibration set utilized to communicate or that has a meaning for a listener.

1.3.1 Originator Agents

According to the originator agents, sound sources are distinguished in at least four categories: *geophonies, biophonies, anthropophonies,* and *technophonies* (Farina et al., 2021a, b) (Fig. 1.1). However, this distinction is not ecologically consistent because natural and technological sounds often produce effects not distinguishable on organisms but remains useful in the field of human evaluation and landscape application. For instance, this categorization is useful to investigate the sonic pollution, to assess the land quality, to create area of well-being and recreation, etc.

Geophonies are produced by thunders, wind, rain, hailstorm, water flow, sea waves, tides, earthquakes, volcanic eruptions, ice cracking, geysers, landsliders, etc. All sounds are the result of a friction between particles. For instance, the sound of wind is the result of friction of air against a substrate (soil, rocks, or water) or vegetation. The sound of a stream is the result of the friction of water against the rocky substrate. The sound of rain is the result of the impact of water droplets with substrates (vegetation, soil, or human infrastructures). Geophonies concur to the characterization of acoustic habitat (Mullet et al., 2017).

Biophonies are produced by specialized organs like vocal cords in mammals, syrinx in birds, vocal sac in frogs, tymbal in cicadas, etc. Not only all the sounds produced by animals during their vocal activity but also the sounds produced by the friction of their bodies against soil, vegetation, water, or air are considered biophonies (e.g., the pawing of herds of ungulates). A special case is the sound produced by the contraction of muscles in some large mammals like horses during very quick movements.

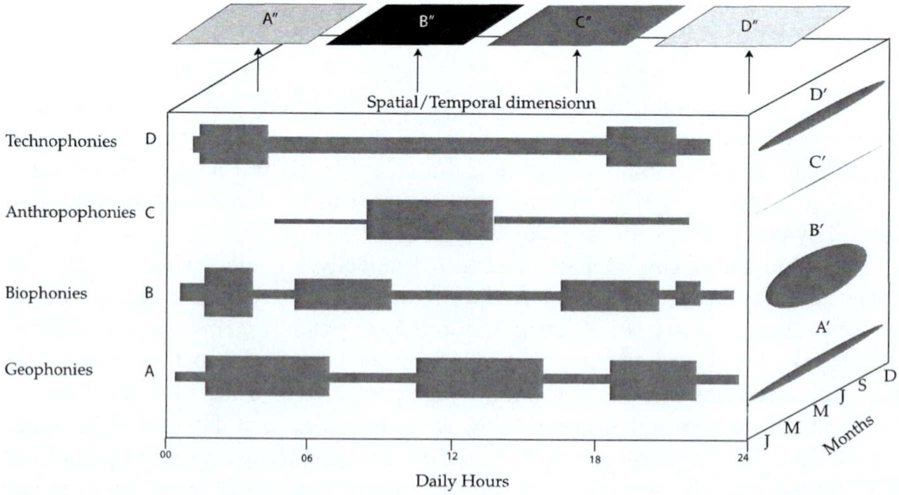

Fig. 1.1 Hypothetical distribution of the four categories of sounds as perceived by humans and their distribution along daily hours (*A:D*), at monthly scale (*A′:D′*), and according to a spatial/temporal dimension (*A″:D″*)

Anthropophonies are the sounds of the human voice (words and whistles), produced by vocal cords and eventually transmitted and magnified by technological devices like radio, computers, tablet, and cell phones.

Technophonies are produced by functioning of static (industrial) or moving machineries like airplanes, cars, trains, boats, tradeship, etc.

1.3.2 Semiosis

According to a semiotic perspective, sounds are distinguished into intentional and unintentional and utilized as a semiotic tool by soniferous and nonsoniferous species to communicate or to explore the surroundings.

Intentional sounds are the result of the broadcasting of vibrations emitted by special organs (e.g., vocal cords, tymbals, etc.) of sonoriferous species or by friction of body and utilized with different purposes (f.i., territory delimitation and patrol, mating, social cohesion, resource tracking, antipredatory function, etc.).

Inside this category, we can make a further distinction in active and passive sounds. Active sound is a sound broadcasted directly by animals by using internal/external organs or part of the body and used to communicate with other individuals or species like bats and cetaceans, to scan the environment for catching preys, or to detect obstacles. Passive sound is every sound produced in the environment and utilized by animals in an eavesdrop attitude after interpretation to complete some functions.

Unintentional sounds are the result of abiotic or biotic matter dynamics and their effects on abiotic (rocks, soil) or biological structures (leaves, animals).

Intentional and unintentional sounds enter into semiotic mechanisms of sensing, interpretation, and reaction utilized by the animals.

1.4 Theoretical Principles in Ecoacoustics

Theoretical principles at the support of ecoacoustics remain scarce (Farina & Gage, 2017) if we exclude some well-recognized bioacoustic theories like the Morphological Adaptation Hypothesis (MAH), the Acoustic Adaptation Hypothesis (AAH), or the Acoustic Niche Hypothesis (ANH). Recently, two new ecoacoustic hypotheses have been added: the Acoustic Community Hypothesis (ACH) (Farina & James, 2016) and the Acoustic Habitat (Mullet et al., 2017).

1.4.1 The Morphological Adaptation Hypothesis (MAH)

The *MAH* recognizes the role of the body size as an important biological constraint of the vocalization organs and their acoustic performance (Wallschläger, 1980). Soniferous species have different organs of vocalization; for instance, in mammals, the sound is emitted by the vocal folds and modulated by the tongue. In songbirds, the sound is produced by the syrinx and modulated by the vocal tract and beak (Fletcher, 2007).

Animals of large size communicate at a greater distance than small animals. For instance, elephants communicate at a greater distance than smaller animals such as wildcats. The same occurs for birds such as green woodpecker (*Picus viridis*) when compared with the distance traveled by the songs of smaller birds like a wren (*Troglodytes troglodytes*).

There are evidences of the relationship between body mass and acoustic frequencies. For instance, Wallschläger (1980) has found in 90 European passerine birds and Fletcher in different species of birds and mammals (Fletcher, 2007) a relationship between acoustic frequencies and body mass expressed by a power law $f \propto M^{0.33}$(f = frequency, M = body mass).

The same evidences have been found by Boeckle et al. (2009) from 76 species of frogs where 25% of the variation in dominant acoustic frequency has been explained by body size.

1.4.2 The Acoustic Adaptation Hypothesis (AAH)

The *AAH* refers to the adaptation of sound emitted by species to environmental characteristics (Morton, 1975; Marten & Marler, 1977; Lemon et al., 1981). There are evidences that species living in dense forests emit sounds with lower frequency (f.i., blackcap (*Sylvia atricapilla*), a small Palearctic songbird) than birds like skylark (*Alauda arvensis*) living in open habitat. This hypothesis has been discussed and confirmed recently by Velásquez et al. (2018) in the four-eyed frog (*Pleurodema thaul*) in eight localities in Chile.

The *AAH* is based on the assumption that sound that decreases at a rate of 6 dB at each doubling of distance for the inverse square law decreases by an excess of attenuation also as the result of several factors such as absorption by air, ground, and vegetation or by processes of reflection and diffraction. We expect that these disturbances vary according to the length of the sound propagation path. Assuming that it remains largely unknown the level of impedance of trees, rocks, grasses, water, and so on, higher frequencies are reflected more by leaves and branches in forests than low frequencies: Local conditions may have a larger effect for sound transmission. Marten and Marler (1977) have found in temperate habitats that animals vocalizing at lower frequency at a height greater than 1 m can transmit at a longer distance their acoustic signals when compared with the same signals emitted close to the soil.

1.4.3 The Acoustic Niche Hypothesis (ANH)

The *ANH* proposes acoustic partitioning mechanisms with which species try to reduce the interspecific frequency overlap (Krause, 1993). This hypothesis, as named by Bernie Krause in 1980 (Krause, 2012), assumes that the sound spectrum is a limited resource and that species try to reduce and minimize acoustic competition. This hypothesis has been proved in frog communities (Duellman & Pyles, 1983) and in cicadas (Sueur, 2002) and is particularly evident in intact environments where a long coexistence between species should have favored an adaptation to reduce competition in sympatric species. This hypothesis integrates the acoustic dimension to the ecological niche concept (Hutchinson, 1957). In disturbed environment where the (acoustic) communities are less organized, the frequency partitioning seems more difficult to be obtained by different species (Malavasi & Farina, 2013). The rapid climate change may have an important impact on the acoustic communities by the change in species composition as consequence of habitat shifts, producing as consequence an increase of acoustic niche competition. There are also evidences that climate has an effect on the acoustic signals of birds song and bat echolocation (Snell-Rood, 2012) and that this change reduces the possibility to species to adapt to the new conditions.

1.4.4 The Acoustic Habitat Hypothesis (AHH)

There are evidences that the quality of the soundscape represents an important criterion in the specific environmental selection. This hypothesis formulated by Mullet et al. (2017) supports the importance of sounds in assuring a good standard of life to species that use such sensed tools.

The acoustic habitat represents the sonic condition that is considered suitable for a species. In particular, noisy environment can represent a deterrent for some species less tolerant to masking sounds.

For a species, the selection of a habitat may be the result of the eavesdropping of sounds broadcasted by other species and interpreted as an indicator of abundance of food resources or of breeding opportunities (Ward & Schlossberg, 2004; Hahn & Silverman, 2007; Ortega, 2012).

The role of information obtained from geophonies comes to the aid of this theory. Wind, rain, water flow, and marine waves become important sources of environmental information. Unfortunately, such sounds often are considered simply noise and discarded from the analysis. We believe that ecoacoustics should reconsider the paradigm of the noise and to start new research on the role of geophonic sounds for the ecology of organisms.

1.4.5 Acoustic Community Hypothesis (ACH)

An acoustic community is defined as the ensemble of biophonic sounds emitted by different species that in some degree impact their behavior and ecology at a particular time and/or at a particular location (Farina & James, 2016).

At difference of the ecological communities, the acoustic communities are ephemeral and are composed only of soniferous species. Several attempts are reported in literature to associate an acoustic community to biodiversity. In some cases, this is possible, like in tropical eco-regions where the level of biophonies is very high and the acoustic partitioning is evident thanks to the contribution of a narrower band acoustic signals of amphibians and insects. In other regions (f.i., in the temperate eco-regions), the level of biophonic emission is modest, and soniferous species, mainly birds, have a song repertoire so varied and rich that the richness of sounds is not correspondent with the richness of species. For instance, in song thrush (*Turdus philomelos*), a bird living in the Palearctic region, males have an extremely rich song repertoire that seems originated by different species at the same time. This effect is really confounding when the sound is analyzed with ecoacoustic metrics, associating such high sonic diversity to different species. Apart from these issues, the concept of acoustic community remains central in the ecoacoustic interpretation of data and in landscape design.

1.5 Developing a Sonic Domain Narrative

Following the Domain Theory as an application of Systems Theories (Hall, 1962; Chestnut, 1967; Andreasen et al., 2014), the sonic domain is the representation of concepts, models, and physical components that regards every typology of sounds. The sonic domain can be explored from an ontological, semiotic, geographical, ecological, and behavioral perspective assuming from time to time different roles (Fig. 1.2).

Table 1.1 defines the main "ecoacoustics elements" belonging to the sonic domain. In this list has been recently added the sonoscape concept that we define as all sonic events occurring in a unit of time or space and at which it has not been applied any specific interpretation. We can say that a sonoscape corresponds to the gradient of energy intercepted by a microphone capsule and sensed but not yet interpreted by an organism.

According to an ontological perspective, we consider the vibroscape as the originator of the sonic domain, and it represents the ensemble of the all energy transmitted in a medium (solid, liquid, or gaseous) by vibrating objects. The vibroscape, for its definition, has no explicit ecological role. To play such role, a vibroscape must be at least sensed by an organism through acoustic and nonacoustic pressure perturbation.

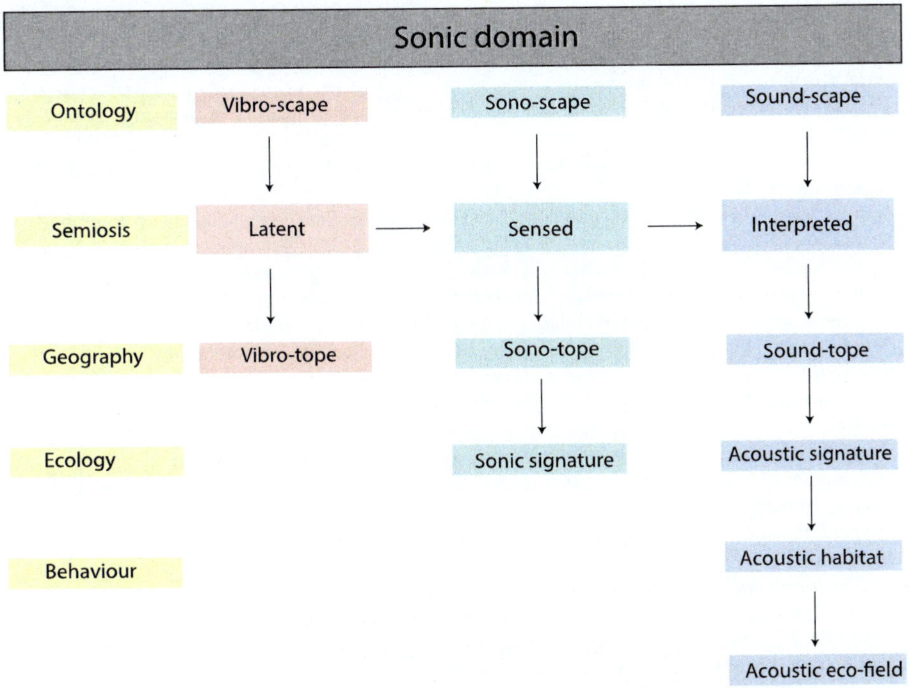

Fig. 1.2 Epistemological representation of the different components of the sonic domain and their relationships according to an ontological, semiotical, geographical, ecological, and behavioral perspective

Sounds are extraordinary vehicles of information and meaning and play a central role for the life of individual species, populations, and communities.

In this narrative, semiosis has a central role to connect the physical phenomenon of the vibration across medium with the ecological and behavioral functions of the organisms.

According to a semiotic perspective, portions of vibroscape from a latent status (no sensed or interpreted) may be considered a *sonoscape* after an organismic sensing (Farina et al., 2021a, b). In turn, a sonoscape after a process of species specific interpretation assumes distinct and functional characters of a soundscape. For instance, the human distinction of sounds in biophonies, geophonies, anthropophonies, and technophonies allows to characterize a soundscape and to assess the quality of an area (f.i., De Coensel & Botteldooren, 2006). For a transitive property, we expect that animals have a similar capacity to distinguish and categorize sounds.

Biophonies and some sounds produced by abiotic dynamics like the fall of water in a fountain can be classified by humans with attributes similar to the ones utilized to evaluate a visual element like a painting or a monument. For instance, a sound can be defined relaxing, inspiring, attractive, etc. The description and classification of sounds as perceived by humans are facilitated by the use of the language that allows to translate the acoustic sensations into words (Davies et al., 2013). This probably does not happen in animals that have not evolved a complex and sophisticate language.

Table 1.1 Some ecoacoustics terminology and their definitions from Farina et al. (2021a, b)

Acoustic community: An aggregation of species that produces sound by using internal or extra-bodily sound-producing mechanisms
Acoustic eco-field: The spatial configuration of a soundscape as a carrier of meaning to track resources
Acoustic habitat: The specific acoustic character of a geographical area selected as favorable by a species for living
Acoustic signature: The sequence of sound that characterizes a species, an acoustic community or soundtope
Alpha biophonic diversity: Biophonic diversity inside an acoustic community
Alpha sonotope diversity: Diversity of ecoacoustic events present in a sonotope
Biophonic diversity: Diversity of sounds produced by an acoustic community
Beta biophonic diversity: Biophonic diversity between two acoustic communities
Beta sonotope diversity: Diversity of ecoacoustic events between two sonotopes
Dawn chorus: The collective acoustic activity of an acoustic community around sunrise
Dusk chorus: The collective acoustic activity of an acoustic community around sunset
Ecoacoustic diversity: Diversity of ecoacoustic events produced by a landscape
Ecoacoustic events: Acoustic signals that after an interpretation have the capacity to influence behavior and ecology of the listeners
Far field: Portion of soundscape that is sensed but does not produce changes in the behaviour of organism
Gamma biophonic diversity: Total biophonic diversity in a landscape
Gamma sonotope diversity: The diversity of ecoacoustic events in a landscape
Near field: Portion of soundscape that produces changes in the behavior of an organism
Noise: Sound that interferes with acoustic communication
Sonoscape: Part of vibroscape sensed by an organism
Sonotone: Area at the edge of soundtopes
Sonotope: A homogenous unit of a soundscape
Soundscape: Part of vibroscape perceived as sound by an organism
Soundtope: A homogeneous unit of soundscape consisting solely of biophonies

In soniferous species, the amount, quality, and loudness of emitted sounds are influenced by the quality of their health conditions (Pedroso et al., 2013), by the amount of food available during the early development (Spencer et al., 2003), and in social organisms, by ranking inside the groups (Wyman et al., 2012). For this, biophonies are efficient biological indicators during field survey and in general for environmental assessment.

In order to reduce interspecific competition, many soniferous species have evolved a strategy of acoustic partitioning based on the selection of preferred frequencies in which to concentrate the acoustic performance. Every soniferous species has a specific *acoustic footprint* that allows their identification from an observer and that evolutionary is the results of a mechanism of *niche partitioning* or *acoustic niche* (Krause, 1993) reducing competition or masking (Fig. 1.3).

Every type of sound, including sounds that people categorize as noise, is a carrier of information for animals, and this has an evolutionary significance. In fact, sounds are fundamental semiotic vehicle able to instantly describe the status of the surrounding. At the difference of the visual cues, that can't represent an interpretable signals when such signal is immobile or when move too fast, often sonic signals trigger unconditional reactions regardless of the condition of an emitter and do not

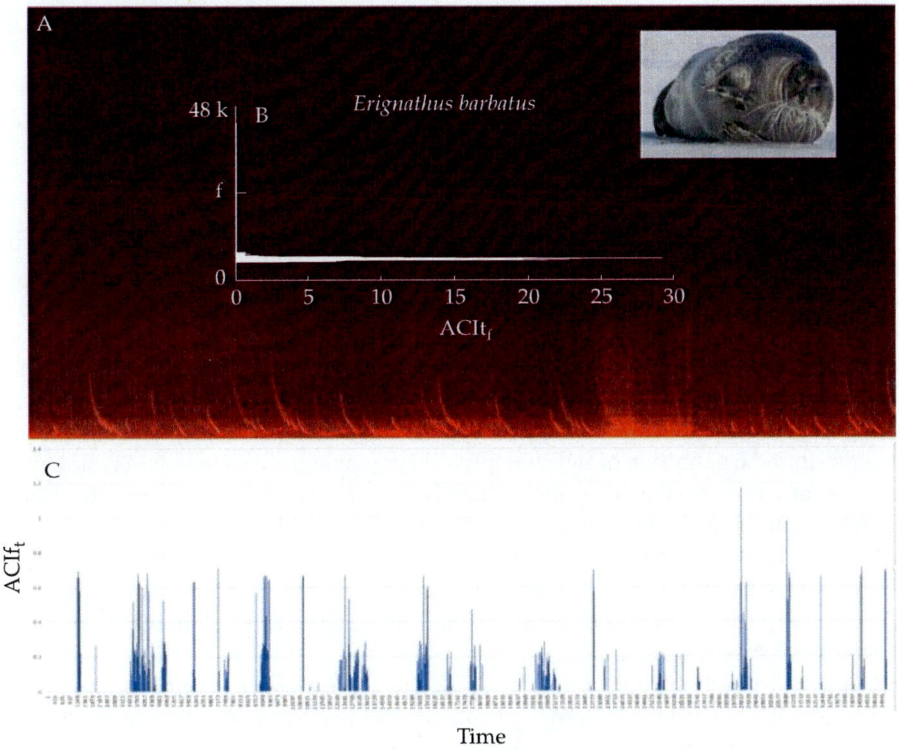

Fig. 1.3 Spectrogram [*A*], Spectral signature ACIt$_f$ [*B*], and temporal signature ACIf$_t$ [*C*] of bearded seal (*Erignathus barbatus*) (courtesy of Giuseppa Buscaino; Parisi et al., 2017). ACIt$_f$ and ACIf$_t$ are two metrics that will be explained later in Chap. 2. Sound is so deep rooted in animal evolution that just a small quantity of sound is sufficient to produce a reaction in a receiver independently by its quality. Acoustic signals above a threshold of amplitude that varies according to the animals are sufficient to alarm or to induce an escaping behavior

require an association with light sources for their identification. For this reason, acoustic signals are particularly used in environments with scarcity or absence of light like deep waters. However, the amplitude of the signal or its distance may create uncertainty in the behavior of a receiver. Distant sounds degrade the information carried and can be ignored by a listener.

According to a geographical perspective, a sonic domain can be distinguished in units called vibrotopes, sonotopes, and soundtopes. These entities represent homogeneous areas for some sonic characters that result from the effect of geographical drivers like soil morphology and aspect, vegetation cover and structure, distance from some vibrational/sonic/acoustic sources, etc.

According to an ecological perspective, a sonic signature is the "phenological" attribute of a sonotope, and it concurs to develop a more precise relationship between the environment and species. The sonic signature is further interpreted by individual species to select an acoustic habitat. Finally, sounds become the driver of important behaviors to track resources and concur to delimit acoustic eco-fields (Farina & Belgrano, 2006).

1.5.1 Sonoscapes and Soundscapes

Today, the possibility to capture sounds is performed by microphonic capsules with the possibility, using digital recorders, to store the acoustic information in a permanent way for successive elaborations. The acoustic files represent the electrical conversion of sounds in a numerical format.

We call the entire set of unclassified/raw sounds "sonoscape" obtained from this acquisition procedure and that is the basis for a successive organismic interpretation obtaining as a result a "soundscape". Figure 1.4 illustrates the sequence from the source until the definition of interpretative agent (sonotopes, soundtopes, sonotones, soundtones, or ecoacoustic events). There is a moment during the survey procedure in which sounds are simply variation in air or water pressure. Then, after the process of interpretation/modeling according to a cultural or an acoustic coding or inferential analysis by using acoustic metrics, sounds become carriers of meanings for species, populations, and communities and are recognized belonging to specific sources and causal phenomena.

The final result obtained with this procedure is the *Soundscape*. A soundscape is defined as the entire set of interpreted sounds obtained by different acoustic sources (*geophonies, biophonies, anthropophonies,* and *technophonies*) as perceived by humans in a specific area/location. This terminology is particularly popular according to a human perspective; however, we have very few information regarding how and at which extension nonhuman organisms cope with and interpret the emerging properties of the sonic phenomena.

There is a conceptual and semantic confusion between a soundscape and the entire set of acoustic vibrations that a microphone can convert in electrical signals. For this, the sound that is transferred, using a recorder into numerical acoustic files, is not coincident with a soundscape but represents a "raw" material that requires a successive interpretation and categorization (Fig. 1.5).

A sonoscape may be represented like a 2D or 3D image after a Fourier Transform (f.i., fast Fourier Transform or FFT procedure) that drives the sound from a temporal domain into a frequential domain arranging a sonic matrix. The spectrogram obtained from this matrix is composed of spectral frequency lines according to the sampling resolution and the size of FFT window adopted.

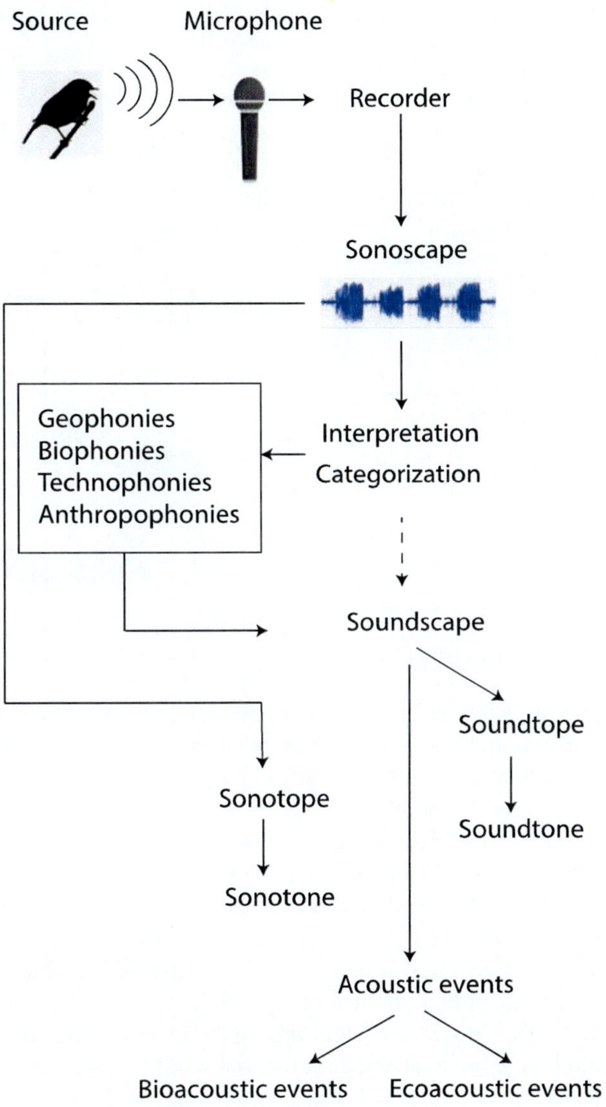

Fig. 1.4 Procedure of capture, storing, transformation, and interpretation of the acoustic signals adopted in ecoacoustics

1.5.2 Sonotopes

Like in a landscape the heterogeneous distribution of land covers creates patches homogeneously at least for some characters, likewise in a sonoscape, the sonic patches have been called by Farina (2014: 17) sonotopes. Sonotopes are the result of a patchy distribution of different sonic sources according to the level and extension of their frequential overlap. The combination of such sources is responsible for the complexity of a sonoscape. At difference of a landscape where the composing

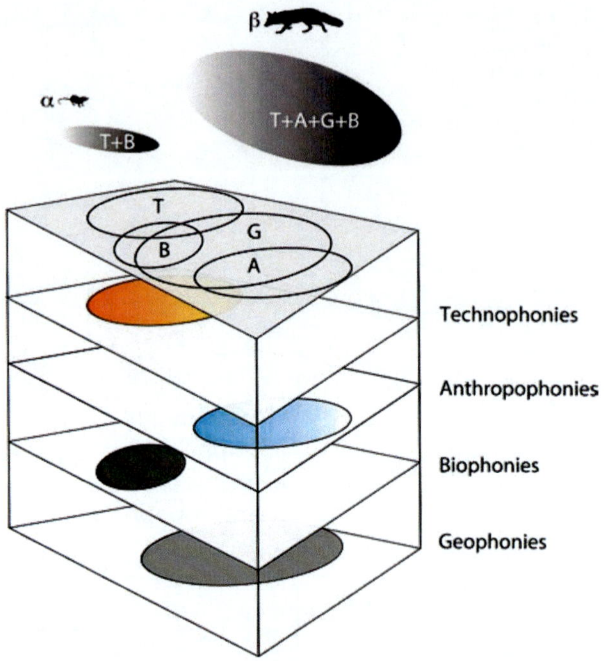

Fig. 1.5 The combination in space of different categories of sonic sources generates distinct sonotopes. Their composition depends on the acoustic capacity of organisms. It is reasonable to admit that small organisms (f.i., shrews) intercept sonotopes composed by a lesser number of sound categories (α) when compared with medium size species that have a larger sensed area (f.i., foxes) (β)

patches show relatively stable characters over a certain period of time, sonoscape presents ephemeral characters in space [s-*sonotope*] and in time [t-*sonotope*].

Defined as a homogeneous unit of a sonoscape, a sonotope can be detected using the Acoustic Complexity Indices (see for more details Chap. 2).

The identification of a sonotope should be coincident with the landscape mosaic. This means that environmental patchiness should match with the differentiated sonotopes. The description of a sonotope is a central theme in ecoacoustic analysis of the landscape. However, very few investigations have been made to relate sonoscapes to landscapes (f.i., Tucker et al., 2014; Fuller et al., 2015).

The identification of sonotopes requires the deployment of a number of recorders at a distance selected according to the characteristic of a landscape. For instance, if the landscape is coarse grained, the recorders will be located more distant than for landscape fine-grained. However, the number of recorders deployed will depend on the level of investigation performed.

There are at least four potential scenarios when we try to couple sonoscapes with landscapes:

(A) The landscape is homogeneous in vegetation patterns and in sonotopes as well.
Explanation: The distribution of soniferous species reflects the distribution of resources.
(B) The landscape is homogeneous in vegetation patterns but results in different sonotopes.

This means that the sonoscape is heterogeneous for causes that cannot be directly and exclusively related to the landcover.

(C) There is a correspondence between characters of a landscape and sonotopes. The distribution of soniferous species reflects the distribution of resources, but unlike the case (A), the patchiness of the landscape overlaps the patchiness of the sonoscape.

(D) Sonotopes are not coincident with landcover. Despite the heterogeneity of the landscape, only one typology of sonotope is observed (Fig. 1.6).

The utilization of sonotope as they emerge from an energy filtering procedure allows to make comparisons between landuse/landcover, soil properties, surface hydrology, aspects, human artifacts, and logistics and sonoscapes.

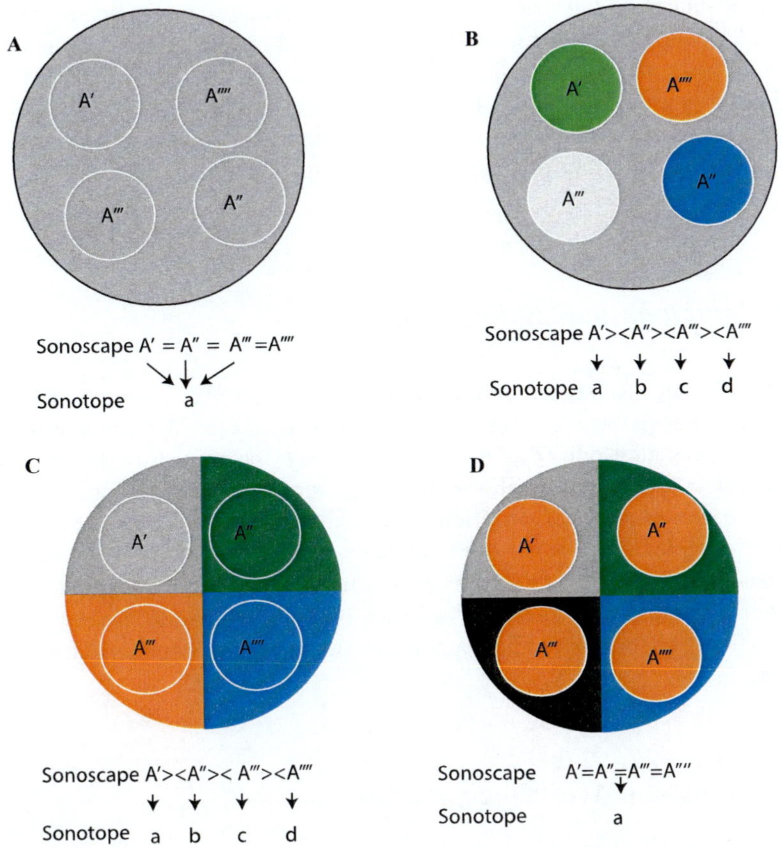

Fig. 1.6 Four possible models in which sonoscape and landscape interact. (**A**) The landscape is homogeneous, and the sonoscape is homogeneous as well producing only one typology of sonotope. (**B**) The landscape is homogeneous, but the sonoscape is different producing four different sonotopes. (**C**) The landscape is composed by four different types of land cover producing as many sonotopes. (**D**) There are not coincidences between landcover and sonoscape. In this case, four different landcovers have the same sonotope. Other combinations are possible, as well

1.5.3 Soundtopes

Soundtopes represent the biophonic patchiness of a soundscape (Farina, 2014: 19). Soundtopes have been defined as a homogeneous unit of soundscape consisting solely of biophonies (Farina et al., 2021a, b). The coordinated ensemble of biophonies can be considered also as an acoustic community, and definitively a soundtope can be considered the "phenotype" of an acoustic community. The spatial variability of a soundtope largely depends on the hearing capacity of the species and by their motility. Species that have a short home range like the birds living in the Mediterranean maquis (f.i., warblers [genus *Sylvia*]) probably will detect only close sounds limiting in this way the potential categories of acoustic sources. Other birds living in open areas like larks (genus *Alauda*, *Lullula*, etc.) emit sounds that spread to longer distances and probably are sensible to a broader range of acoustic signals. Temporal modulation is the major character of biophonies that show cyclic dynamics. For instance, morning choruses in birds are examples of such recurrent biophonies (Farina et al., 2021a, b). Other sonic categories like the technophonies have some diel recurrences connected with the social and economic model of production of goods and services but in general have a simpler internal structure when compared with biophonies.

1.5.4 Sonotones and Soundtones

Like in landscape ecology where the areas of contact between two or more patches are called ecotones from which emerge particular attributes like a higher diversity or a higher predation rate (Risser, 1995), similarly, the contact between different sonotopes and soundtopes creates respectively sonotones and soundtones (Farina, 2014: 19; Farina et al., 2021a, b). Sonotones and soundtones are areas of particular acoustic complexity where the acoustic signals coming from different sonotopes/soundtopes blend together increasing the acoustic uncertainty. This may have effects on soniferous species when such borders are crossed. Investigations of the real effects of sonotones and soundtones are strongly recommended for future research.

1.5.5 Sonic and Acoustic Signature

We define sonic signature as the ecological footprint or sonic expression of a sonotope. The geographical peculiarities of a landscape represent drivers, constraints, or regulation for the sonic fields generated by whatsoever sources (natural or anthropic). Every portion of a landscape is embedded into a specific sonotope that may be variable in space, time, sources, and intensity. Such variability operates at different temporal and spatial scale and is sensed differently according to the species.

A sonic signature can be the result of a single sound source (f.i., rain, wind, road traffic) or more often by the combinations of different sources. Its impact on the ecology and behavior of species is species-specific. It is possible to distinguish a frequential and a temporal sonic signature.

The frequential sonic signature is the direct expression of $ACIt_f$ obtained by summing all the $ACIt_f$ values along a temporal sequence distinct in every spectral line (see Chap. 2). The temporal sonic signature is obtained from the aggregated values of $ACIf_t$

The sonic signatures can be used to compare the behavior of a sonoscape under different environmental conditions (e.g., sunny, windy, rainy days, anthropogenic disturbance, etc.) and along hourly, daily, or seasonal periods.

The Acoustic signature is defined as a sequence of sounds that characterizes a species or an acoustic community. Acoustic signature is a term to describe the acoustic footprint along biophonic frequencies and like for the sonic signature can be distinguished into a spectral acoustic signature and a temporal acoustic signature. This signature can be obtained by applying ACI metrics ($ACIt_f$) to specific spectral line that represent biophonic expressions.

1.5.6 The Ecoacoustic Events

The ecoacoustic event (EE) represents a model in which the combination of three ACI metrics ($ACIf_t$, $ACIf_t$ evenness, and $ACIt_f$ evenness) facilitates the comprehension of the complex dynamics of the sonoscapes (Fig. 1.7). Extensive details about the way to calculate the EE are reported in Chap. 2.

The term “event” has either a semiotic meaning of an interpreted sonic phenomenon, like a thunder, a bell, or a song, or a statistical pattern meaning. While the first category belongs to distinguished and recognized events and is easy to be categorized, the second requires more explanations.

To better distinguish these two typologies, we introduce the term Ecoacoustic Events of Sensed and Interpreted Sonic Phenomena (EE_{sisp}) for the first category and Ecoacoustic Events of Statistical Patterns Sonic Phenomena (EE_{spsp}) for the second category.

EE_{sisp} may be recognized using a reference system. For instance, a wind, a wave, or sea movement.

EE_{spsp} is used to advancing the interpretation of sonic phenomena and to investigate sonic patterns and processes searching relations with other environmental factors.

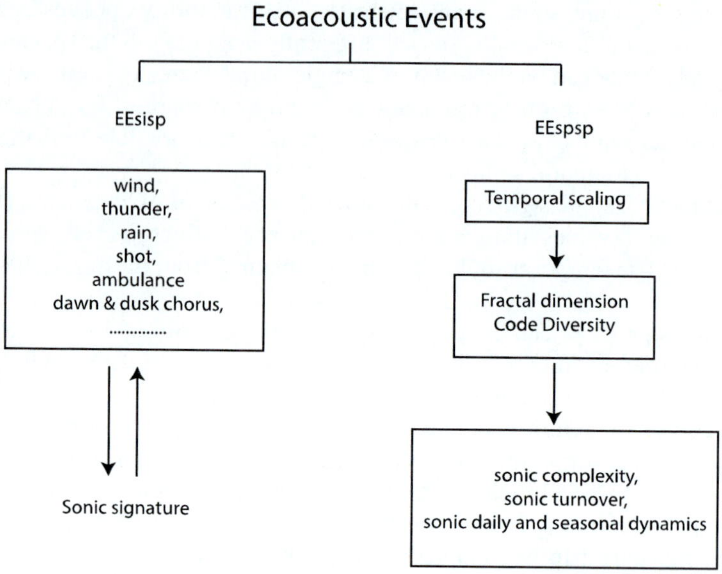

Fig. 1.7 Conceptual representation of the two typologies of Ecoacoustic Events: the Ecoacoustic Events of Sensed and Interpreted Sonic Phenomena (EE_{sisp}) and the Ecoacoustic Events of Statistical Patterns Sonic Phenomena (EE_{spsp}) that adopting a multiscale approach allow to calculate the fractal dimension and code richness, entropy, and temporal distribution. These parameters can be used to assess sonic complexity, sonic turnover, and sonic daily and/or seasonal dynamics

1.5.7 Sonic Matrix

We introduce the concept of sonic matrix or sound spectrum as the result of a Fourier Transform of an acoustic file. A sonic matrix is composed with n spectral lines (according to the level of sampling adopted) and m columns according to the temporal duration of the acoustic file selected or of the portion of the same file analyzed. In our representation, n frequency categories and m the time intervals.

Figure 1.8 represents, two theoretical matrices, one empty (no signals at all) and one completely full of signals with the same amplitude, a condition that is not easy to be found in nature (complete silence, complete soundy) but that can be helpful to explaining the concept of sonic matrix. The amount of information in an empty matrix is not too dissimilar from the information present into a full matrix because in both the cases novelties are absent. This may represent a paradox, but if we consider information as the difference between distinct characters (Bateson, 1970), all the values of a full matrix have a total information equal to an empty matrix. Figure 1.9 describes a real example of a spectrogram full of acoustic signals originated in a tropical forest.

All the metrics presented in this book process data resulting from a sonic matrix.

Between some of the several possible patterns observable in a sonic matrix, we have represented two simple patterns: respectively a horizontal signal that pertains a unique sequence, many sounds of insects have this structure and a vertical signal that crosses all the frequencies (an example is from alarm calls of birds) (Fig. 1.10).

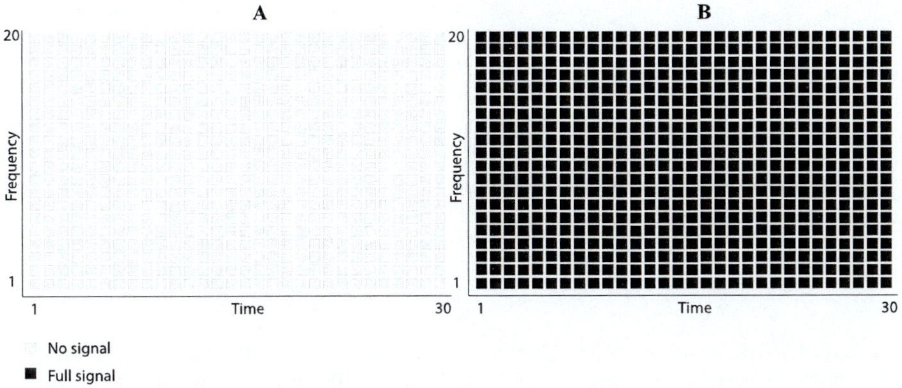

Fig. 1.8 Example of sonic matrix, in (**A**) completely empty (no signals) and in (**B**) completely full of signals. Two conditions that in nature are very improbable. Every discrete signal is represented by a square with coordinates *x* and *y* and by a color according to the amplitude of the signal

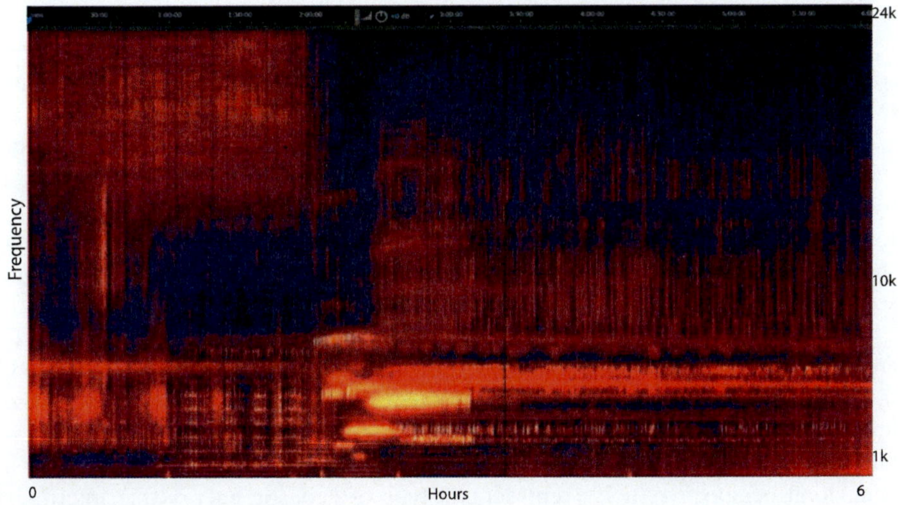

Fig. 1.9 A sonic matrix obtained from a 6 h of continuous recording in a tropical forest (Borneo) (courtesy of David Monacchi). From this image it is possible to note the presence of biophonies in every spectral line. Most of the sounds is mainly produced by insects, frogs, and birds, and finally by mammals. The high density of sounds reflects the huge biodiversity of this region. From this picture it is also possible to observe the temporal turnover of species and their spectral partitioning

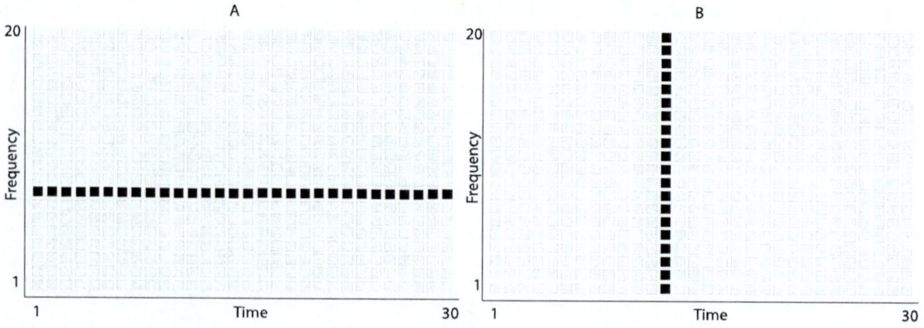

Fig. 1.10 (**A**) This model that can be applied to many sounds of insects that have a signal restricted to a narrow range of frequencies and persist for some time. (**B**) This model is characterized by an event occurring along the entire set of frequencies for a short temporal window [e.g., the alarming call of the European robin (*Erithacus rubecula*)]

1.6 Sounds Interpreted as Noise

We dedicate a special attention to the acoustic noise for the wide implication that this sound has on human life, reserving surprise in many aspects of the ecology of soniferous species.

Noise is a physical property of matter and energy described in many fields of the scientific research from statistics to seismology, optics, and biological time series (e.g., Scales & Snieder, 1998; Ideker et al., 2011).

In acoustics, noise is a sound characterized by poor information (high level of vibrational disorder) that masks other sounds and that affects the active space used by terrestrial and aquatic animals for acoustic communication (f.i., Naguib, 2013, for a revision). The term refers to an unwanted sound, or a disturbing or masking sound. It is defined also as "An acoustic background of incoherent, or random sound" (Farina et al., 2021a, b).

In acoustic physics, noise represents a random oscillation in a signal with different spectral density being described by different colors (f.i., white, pink, brown, red, blue, violet). The signal-to-noise ratio represents the level of a signal relative to the level of background noise. In ecoacoustics, noise is a category of sound that produces effects on the life of animals and on their behavior. For instance, from a behavioral perspective, noise represents a type of acoustic energy that interferes with the communication between an emitter and a particular receiver (Luther & Gentry, 2013), and such interference often masks or impedes communication between organisms (Truax, 1999).

Noise has received a great interest from investigators especially in relationship to animal behavior (f.i., Brumm & Slabbekoorn, 2005; Habib et al., 2007) and community ecology (Francis et al., 2009), for a review (Barber et al., 2009; Nowacek et al., 2007; Slabbekoorn et al., 2010).

Noises of anthropogenic origin may have consequences on individual species, and this noise is of growing concern (Chan et al., 2010) for people as well (Babisch et al., 2005; Barber et al., 2009). We have few information on the animal capacity to interpret the noise like human doing; however in ecology, there are evidences that noise is utilized by species to assess their acoustic habitat (Mullet et al., 2017) and that some species avoid noisy habitats (Francis et al., 2009). The presence of anthropogenic noise may change the biogeography of species, at least at local scale. There are noise-tolerant species that occupy area with high level of noise and that substitute species less tolerant (Francis et al., 2009). The noise is a phenomenon that remains little known and questionable, but is accepted that noise is a portion of sound that concurs to improve the knowledge of the surrounding and the acoustic habitat evaluation. Noise is a category of sounds that ultimately contributes to the environmental information not for its internal structure (see the biophonies) but for its occurence and intensity variation in time. We associate noise to geophonies, technophonies, and anthropophonies but also a breeding colony of birds like penguins or other sea birds like Cape gannet (*Morus capensis*) or honey bee colonies may create a very noise environment.

Different strategies can be used by animals like birds to reduce the masking or disturbing effect of noise by changing the amplitude of the acoustic signal, known as Lombard effect (Lombard, 1911), changing frequency (Narins et al., 2004), signal redundancy (Potash, 1972; Wiley, 1994; Brumm & Slater, 2006; Díaz et al., 2011), and behavior (Lupo et al., 1991; Rabin et al., 2006; Chan et al., 2010).

Geophonies that are fundamental components of the vivo-scape species specific (Farina & James, 2021) concur to the habitat selection of species. The dynamic of geophonies are poorly investigated and controversial. Probably in animals that have a great capacity to move from one area to another, like swifts and petrels, geophonies may represent important cues to decide the more favorable area in which to stay. Strong winds may orient these species to move in more acoustically calm areas where to find food. It is also reasonable, due to the great sensitivity of migrating birds to climatic events, that geophonies could be used as cues to navigate in the safe direction. There are growing evidence that the infrasounds produced by tornado can displace breeding birds like the golden-winged warbler (*Vermivora chrysoptera*) toward more far areas avoiding the effect of this extreme event (Elbing et al., 2018). The capacity to hear infrasounds is well known in mammals like elephants and is hypothesized also in birds more than believed until now where the infrasound perception may have tympanic and extratympanic pathways (Zeyl et al., 2020). If the range of acoustic and nonacoustic variation of air-borne pressure perturbation is detected by several animals, the investigation of a geophonic signature across a landscape could become a relevant element of investigation confirming the importance of the sonic environment in ecology and behavior.

1.7 Sonic Information Systems

The use of sounds to investigate the relationship between landscape and sonoscape dates back some years. The spatiality of a landscape and the geography of sounds have recently attracted scholars in human ecology, geographers (Pocock, 1989; Chitra et al., 2020; Fan et al., 2021), and ecologists (Farina & Pieretti, 2014).

Sonotopes become the geographical units homogeneous for some sonic characters, and their mapping obtained after the deployment of a grid of microphones represents the basis of a Sonic Information System.

It is evident that the distribution of vegetation and more in general the structure of the land mosaic are perceived with species-specific modalities. According to the habitat preferences, species are consequently distributed along a precise land cover gradient. Direct observation and capture–recapture methods allow both to estimate the presence of species. The recent utilization of sound (biophonies) results very promising (Fuller et al., 2015); however, it is necessary to consider that sound has no precise border and that its intensity is reduced of 6 dB doubling the distance from the source.

To estimate the distance at which a sound is produced, it is essential to cope sound and the landcover of a landscape. For instance, in a recent research, Farina (2018) has utilized a close distance of only 30 m to detect species and differentiate the fine-grained mosaic of a Mediterranean rural landscape.

The spatial comparison between sonoscapes and landscapes requires the interpretation of sonoscapes as a collection of sonotopes delimited by a sonic energy gradient after the application of energy filter procedures.

For this reason, we advance a model to couple sonoscape and landscape based on the approaches extensively utilized by the *Geographic Information Systems* (GIS).

We call such system *Sonic Information System* (SIS), a system that can manage sonic raw data and data processed by different metrics according to different spatial/temporal resolution. SIS is based on the spatial arrangement of *recording stations* (RS) posed at a regular distance and formally represented by a hexagonal grid (Birch et al., 2007). The addition of RS may create fractal-like/hierarchical hexagons of different size (Fig. 1.11).

In the cases described here, we have adopted a hexagon of 200 x 200 m of dimension, but this dimension can be changed accordingly.

The minimum deployment of recording stations (RS) is based on 7 units covering an area of 600 × 600 m, but this limit is imposed by the costs of devices, and by the time requested to handle recorders and to process data. Adding other 12 recorders, we cover an area of 1000 × 1000 m totaling 19 recorders. In the exercise in Chap. 4 are reported examples of an SIS composed of 11 recorders deployed at a distance of 200 m.

After the deployment of the recorders, it is necessary to fix the energy threshold categories to delimit the sonotopes. The number of categories of sonotopes should be fixed empirically (f.i., three or five categories) estimating the distance at least of the more far categories (faintest sounds).

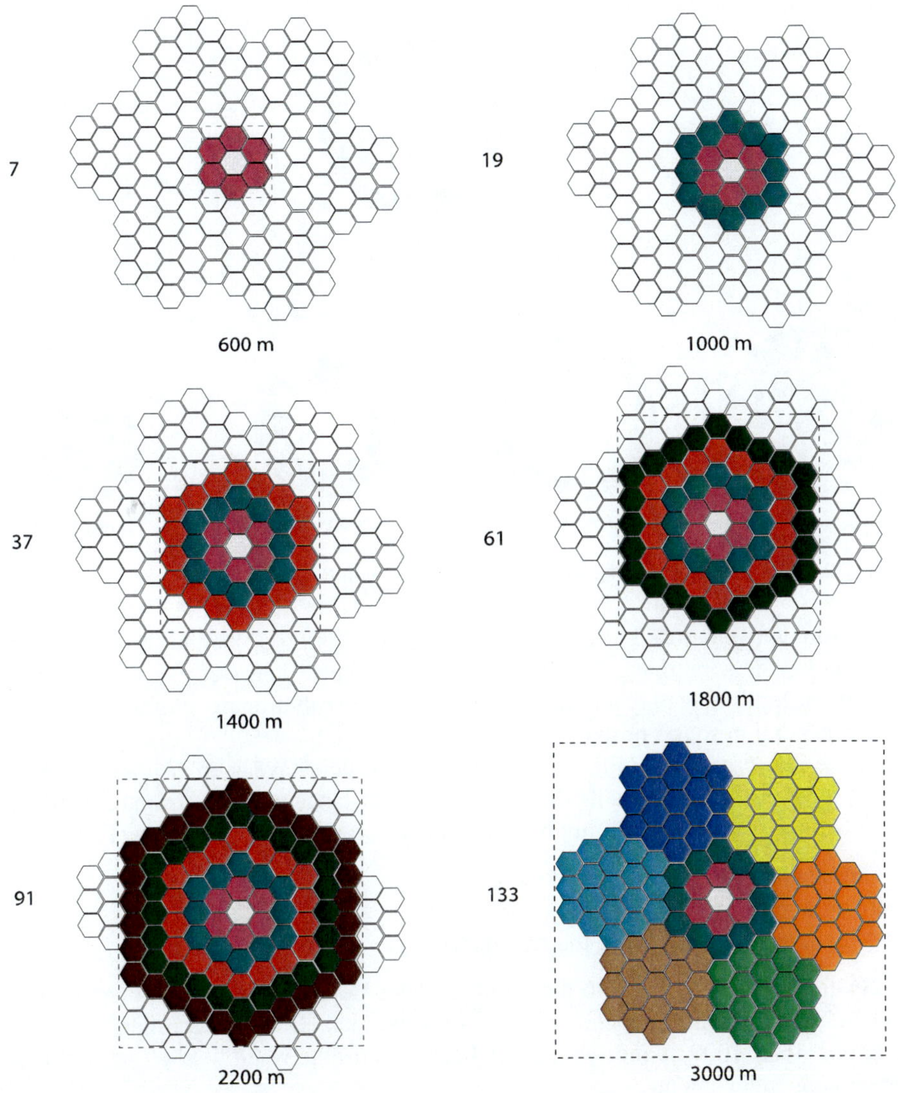

Fig. 1.11 Representation of a potential Sonic Information System (SIS) according to a hexagonal grid and based on the same principles that guide the Geographical Information Systems. Number of recorders and the linear distance of the covered area are reported

Then using an SIS that operates with the logic of a true Geographic Information System (GIS) is possible to produce maps of land cover/land use, and tracing circular plot from the center of the recording station (Fig. 1.12). A correlation analysis may assist us to evaluate the relationship between sonotope categories and landscape mosaic.

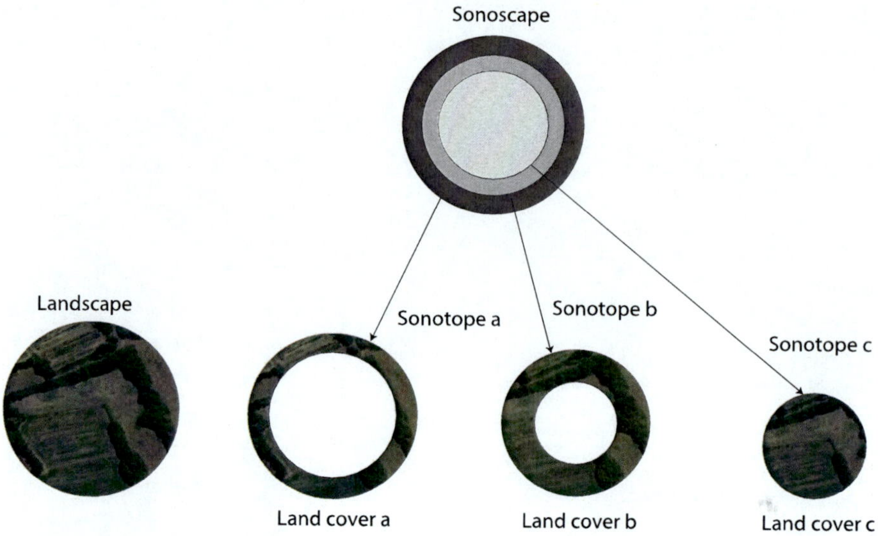

Fig. 1.12 Schematic representation of sonotope overlap with land mosaic

GIS may also be used to calculate the distance from objects and their potential impacts on the sonotopes. For instance, the soil aspect and steepness of hill and mountain slopes may play an important role in the distribution of soniferous species and in general in sound propagation.

In the case proposed in Fig. 1.13, 19 digital recorders have been disposed according to a hexagonal grid, posing the recorders at a distance of 200 m from each other.

A SIS can be developed into three dimensions, represented, respectively, by:

(a) Spatial distribution of digital recorders
(b) Temporal resolution of data processing
(c) Temporal aggregation of the ACI metrics

Spatial resolution refers to the number of RS deployed in the field (minimum 7–19). The upper limit depends on the size of project. Nineteen recorders cover an area of 1000 × 1000 m and seem a good compromise between field effort and data processing engagement.

The temporal resolution refers to time scale at which the raw data are processing using ACI metrics. For instance, 1 s, 2 s, etc.

The temporal aggregation is the time interval at which data are aggregated. Usually, we can adopt different temporal resolution to observe phenomena like dawn or dusk choruses, or to examine the trend of metrics at hour, half day, day-night, daily, monthly, seasonally, etc.

A spatial grid is created according to the number of recorders and represents a reference system to compare land use or other thematic maps with the characters of sonotopes. This grid can be filled with the ACI metrics aggregated per recording stations. A number of spatial maps can be created according to the different

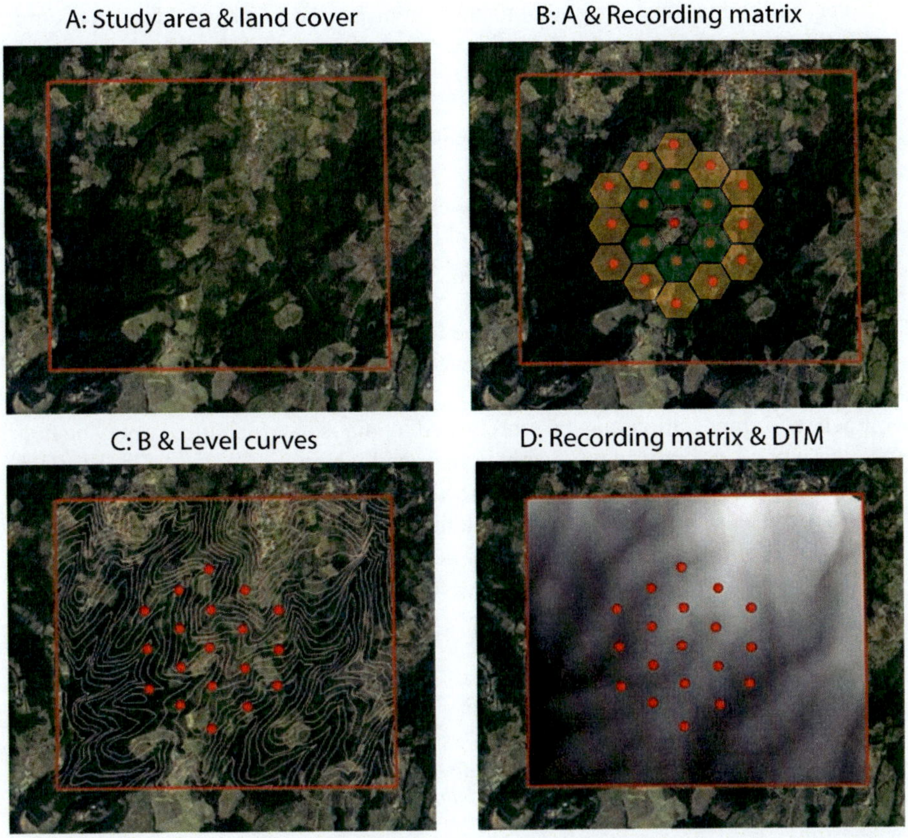

Fig. 1.13 A sonic information system SIS is composed by a spatial deployment of recorders, in this case 19 digital recorders, disposed according a hexagonal net. The SIS becomes an integration between a GIS and sonic information transformed into a spatial realm thanks to the use of threshold belts. *DTM* digital terrain model. In this example, the study site of Magliano (44° 14′ N, 10°04′ E)

temporal aggregation. For instance, we can have 12 maps for every hour of the day, or four maps according to the seasons. It is evident that each map is the result of an aggregation of data.

A second matrix with dimension 19 × 24 is composed of a coordinate system based on the number of recorders (19) and the hourly aggregation of ACI metrics ($ACIf_t$, $ACIt_f$, EE, SSD). A third matrix with dimension 19 × 30 is composed by a coordinate system based on the number of recorders (19) and by the number of days. A fourth matrix with dimension 19 × 4 is composed by a coordinate system based on the number of recorders (19) and by the seasons. Other temporal scaling is possible, as well.

Each temporal aggregation is repeated according to the ACI metrics (4) and the energy threshold utilized (1, 2, 3, 4, …).

Every individual recording station can be represented in a spatial format according to the temporal resolution adopted and the categories of ACI metrics values. We can plot frequency categories of $ACIt_f$ (e.g., with belts of 100 Hz) with a temporal resolution.

The entire set of matrices can be developed according to the processing scale from 1 s onward.

We have to clarify that the spatial dimension is represented by the distribution of recorders (f.i., 19) and the energy threshold utilized (f.i., 1,2,3, … belts).

The temporal dimension is represented by the aggregation of data according to the temporal scale of interest.

A process of interpolation can help the best visualization of data.

In summary, an SIS is composed of several matrices of data rendered in a spatial format (maps).

Each recording station may be represented by maps in which the *x,y* may be hour x day or day x season, or day(season) x year, and where z = $ACIf_t$, $ACIt_f$, EE (temporal distribution, richness, entropy), or SSD (Sonic Signature Dissimilarity).

When the entire set of recording stations is represented, different maps are obtained inserting every metric in the *x,y* geographic coordinates.

References

Agranat, I. (2007). *Automatic detection of caeruleus warbler using autonomous recording units and song scope bioacoustic software*. Wildlife Acoustics.

Agranat, I. (2009). *Automatically identifying animal species from their vocalizations*. Wildlife Acoustic.

Andreasen, M. M., Howard, T. J., & Bruun, H. P. L. (2014). Domain theory, its models and concepts. In *An anthology of theories and models of design* (pp. 173–195). Springer.

Babisch, W., Beule, B., Schust, M., Kersten, N., & Ising, H. (2005). Traffic noise and risk of myocardial infarction. *Epidemiology, 16*, 33–40.

Barber, J. R., Crooks, K. R., & Fristrup, K. M. (2009). The costs of chronic noise exposure for terrestrial organisms. *Trends in Ecology and Evolution, 25*(3), 180–189.

Bateson, G. (1970). Form, substance, and difference. *General Semantic Bulletin, 37*, 5–13.

Birch, C. P., Oom, S. P., & Beecham, J. A. (2007). Rectangular and hexagonal grids used for observation, experiment and simulation in ecology. *Ecological Modelling, 206*(3–4), 347–359.

Boeckle, M., Preninger, D., & Hödl, W. (2009). Communication in noisy environments. I: Acoustic signals of *Staurois latopalmatus* Boulenger 1887. *Herpetologica, 65*(2), 154–165.

Brumm, H., & Slabbekoorn, H. (2005). Acoustic communication in noise. In P. J. B. Slater, C. T. Snowdon, T. J. Roper, H. J. Brockmann, & M. Naguib (Eds.), *Advances in the study of behavior* (Vol. 35, pp. 151–209). Academic Press.

Brumm, H., & Slater, P. J. (2006). Ambient noise, motor fatigue, and serial redundancy in chaffinch song. *Behavioral Ecology and Sociobiology, 60*(4), 475–481.

Chan, A. A. Y. H., Giraldo-Perez, P., Smith, S., & Blumstein, D. T. (2010). Anthropogenic noise affects risk assessment and attention: The distracted prey hypothesis. *Biology Letters, 6*(4), 458–461.

Chestnut, H. (1967). *Systems engineering methods*. Wiley.

Chitra, B., Jain, M., & Chundelli, F. A. (2020). Understanding the soundscape environment of an urban park through landscape elements. *Environmental Technology & Innovation, 19*, 100998.

Davies, W. J., Adams, M. D., Bruce, N. S., Cain, R., Carlyle, A., Cusack, P., Hall, D. A., Hume, K. I., Irwin, A., Jennings, P., Marselle, M., Plack, C. J., & Poxon, J. (2013). Perception of soundscapes: An interdisciplinary approach. *Applied Acoustics, 74*(2), 224–231.

De Coensel, B., & Botteldooren, D. (2006). The quiet rural soundscape and how to characterize it. *Acta Acustica United with Acustica, 92*(6), 887–897.

Depraetere, M., Pavoine, S., Jiguet, F., Gasc, A., Duvail, S., & Sueur, J. (2012). Monitoring animal diversity using acoustic indices: Implementation in a temperate woodland. *Ecological Indicators, 13*(1), 46–54.

Díaz, M., Parra, A., & Gallardo, C. (2011). Serins respond to anthropogenic noise by increasing vocal activity. *Behavioral Ecology, 22*(2), 332–336.

Duellman, W. E., & Pyles, R. A. (1983). Acoustic resource partitioning in anuran communities. *Copeia, 1983*, 639–649.

Elbing, B. R., Petrin, C., & Van Den Broeke, M. S. (2018). Monitoring infrasound from a tornado in Oklahoma. *The Journal of the Acoustical Society of America, 143*(3), 1808–1808.

Fan, Q. D., He, Y. J., & Hu, L. (2021). Soundscape evaluation and construction strategy of park road. *Applied Acoustics, 174*, 107685.

Farina, A. (2014). *Soundscape ecology: Principles, patterns, methods and applications.* Springer.

Farina, A. (2018). Perspectives in ecoacoustics: A contribution to defining a discipline. *Journal of Ecoacoustics, 2*, TRZD5I.

Farina, A., & Belgrano, A. (2006). The eco-field hypothesis: Toward a cognitive landscape. *Landscape Ecology, 21*(1), 5–17.

Farina, A., & Gage, S. H. (2017). Ecoacoustics: A new science. In A. Farina & S. H. Gage (Eds.), *Ecoacoustics: The ecological role of sounds* (pp. 1–9). Wiley.

Farina, A., & James, P. (2016). The acoustic communities: Definition, description and ecological role. *Biosystems*. https://doi.org/10.1016/j.biosystems.2016.05.011

Farina, A., & James, P. (2021). Vivoscapes: An ecosemiotic contribution to the ecological theory. *Biosemiotics*, 1–13. https://doi.org/10.1007/s12304-021-09406-2

Farina, A., & Pieretti, N. (2014). Sonic environment and vegetation structure: A methodological approach for a soundscape analysis of a Mediterranean maqui. *Ecological Informatics, 21*, 120–132.

Farina, A., Gage, S. H., & Salutari, P. (2018). Testing the ecoacoustics event detection and identification (EEDI) approach on Mediterranean soundscapes. *Ecological Indicators, 85*, 698–715.

Farina, A., Eldridge, A., & Li, P. (2021a). Ecoacoustics and multispecies semiosis: Naming, semantics, semiotic characteristics, and competencies. *Biosemiotics*. https://doi.org/10.1007/s12304-021-09402-6

Farina, A., Righini, R., Fuller, S., Li, P., & Pavan, G. (2021b). Acoustic complexity indices reveal the acoustic communities of the old-growth Mediterranean forest of Sasso Fratino integral natural reserve (Central Italy). *Ecological Indicators, 120*, 106927.

Fletcher, N. H. (2007). Animal bioacoustics. In T. D. Rossing (Ed.), *Springer handbook of acoustics* (pp. 785–802). Springer.

Francis, C. D., Ortega, C. P., & Cruz, A. (2009). Noise pollution changes avian communities and species interactions. *Current Biology, 19*(16), 1415–1419.

Fuller, S., Axel, A. C., Tucker, D., & Gage, S. H. (2015). Connecting soundscape to landscape: Which acoustic index best describes landscape configuration? *Ecological Indicators, 58*, 207–215.

Habib, L., Bayne, E. M., & Boutin, S. (2007). Chronic industrial noise affects pairing success and age structure of ovenbirds Seiurus aurocapilla. *Journal of Applied Ecology, 44*(1), 176–184.

Hahn, B. A., & Silverman, E. D. (2007). Managing breeding forest songbirds with conspecific song playbacks. *Animal Conservation, 10*(4), 436–441.

Hall, A. D. (1962). *A methodology for systems engineering*. Van Nostrand.

Hill, A. P., Prince, P., Snaddon, J. L., Doncaster, C. P., & Rogers, A. (2019). AudioMoth: A low-cost acoustic device for monitoring biodiversity and the environment. *HardwareX, 6*, e00073.

Hutchinson, G. E. (1957). Concluding remarks. *Cold Spring Harbor Symposia on Quantitative Biology, 22*, 415–427. GS SEARCH.

Ideker, T., Dutkowski, J., & Hood, L. (2011). Boosting signal-to-noise in complex biology: Prior knowledge is power. *Cell, 144*(6), 860–863.

Krause, B. (1993). *The niche hypothesis, a hidden symphony of animal sounds, the origins of musical expression and the health of habitats* (pp. 156–160). The Explorers journal.

Krause, B. (2012). *The great animal orchestra*. Little, Brown.

Krause, B., & Farina, A. (2016). Using ecoacoustic methods to survey the impacts of climate change on biodiversity. *Biological Conservation, 195*, 245–254.

Lemon, R. E., Struger, J., Lechowicz, M. J., & Norman, R. F. (1981). Song features and singing heights of American warblers. Maximization or optimization of distance? *The Journal of the Acoustical Society of America, 69*(4), 1169–1176.

Lombard, E. (1911). Le signe de l'elevation de la voix. *Annales des Maladies de L'Oreille et du Larynx, 37*, 101–199.

Lupo, C., Lodi, L., Paluffi, G., & Viti, A. (1991). Central and peripheral endocrine correlates of the immobility reaction in the toad Bufo bufo. *Behavioural Processes, 24*(1), 1–7.

Luther, D., & Gentry, K. (2013). Sources of background noise and their influence on vertebrate acoustic communication. *Behaviour, 150*, 1045–1068.

Malavasi, R., & Farina, A. (2013). Neighbours' talk: Interspecific choruses among songbirds. *Bioacoustics, 22*(1), 33–48.

Marten, K., & Marler, P. (1977). Sound transmission and its significance for animal vocalization. *Behavioral Ecology and Sociobiology, 2*, 271–290.

Morton, E. (1975). Ecological sources of selection on avian sounds. *The American Naturalist, 109*(965), 17–34.

Mullet, T. C., Farina, A., & Gage, S. H. (2017). The acoustic habitat hypothesis: An ecoacoustics perspective on species habitat selection. *Biosemiotics, 10*(3), 319–336.

Naguib, M. (2013). Living in a noisy world: Indirect effects of noise on animal communication. *Behaviour, 150*(9–10), 1069–1084.

Narins, P. M., Feng, A. S., Lin, W., Schnitzler, H. U., Denzinger, A., Suthers, R. A., & Xu, C. (2004). Old World frog and bird vocalizations contain prominent ultrasonic harmonics. *The Journal of the Acoustical Society of America, 115*(2), 910–913.

Nowacek, D. P., Thorne, L. H., Johnston, D. W., & Tyack, P. L. (2007). Responses of cetaceans to anthropogenic noise. *Mammal Review, 37*, 81–115.

Ortega, C. P. (2012). Chapter 2: Effects of noise pollution on birds: A brief review of our knowledge. *Ornithological Monographs, 74*(1), 6–22.

Parisi, I., De Vincenzi, G., Torri, M., Papale, E., Mazzola, S., Bonanno, A., & Buscaino, G. (2017). Underwater vocal complexity of Arctic seal Erignathus barbatus in Kongsfjorden (Svalbard). *The Journal of the Acoustical Society of America, 142*(5), 3104–3115.

Pedroso, S. S., Barber, I., Svensson, O., Fonseca, P. J., & Amorim, M. C. P. (2013). Courtship sounds advertise species identity and male quality in sympatric Pomatoschistus spp. gobies. *PLoS One, 8*(6), e64620.

Pieretti, N., Farina, A., & Morri, D. (2011). A new methodology to infer the singing activity of an avian community: The Acoustic Complexity Index (ACI). *Ecological Indicators, 11*(3), 868–873.

Pijanowski, B. C., Villanueva-Rivera, L. J., Dumyahn, S. L., Farina, A., Krause, B., Napoletano, B. M., Gage, S. H., & Pieretti, N. (2011a). Soundscape ecology: The science of sound in the landscape. *Bioscience, 61*(3), 203–216.

Pijanowski, B. C., Farina, A., Dumyahn, S. L., Krause, B. L., & Gage, S. H. (2011b). What is soundscape ecology? *Landscape Ecology, 26*(9), 1213–1232.

Pocock, D. (1989). Sound and the geographer. *Geography, 74*(3), 193–200.

Potash, L. M. (1972). Noise-induced changes in calls of the Japanese quail. *Psychonomic Science, 26*(5), 252–254.

Rabin, L. A., Coss, R. G., & Owings, D. H. (2006). The effects of wind turbines on antipredator behavior in California ground squirrels (Spermophilus beecheyi). *Biological Conservation, 131*(3), 410–420.

Risser, P. G. (1995). The status of the science examining ecotones. *Bioscience, 45*(5), 318–325.

Ross, S. R. J., Friedman, N. R., Yoshimura, M., Yoshida, T., Donohue, I., & Economo, E. P. (2021). Utility of acoustic indices for ecological monitoring in complex sonic environments. *Ecological Indicators, 121*, 107114.

Scales, J. A., & Snieder, R. (1998). What is noise? *Geophysics, 63*(4), 1122–1124.

Slabbekoorn, H., Bouton, N., van Opzeeland, I., et al. (2010). A noisy spring: The impact of globally rising underwater sound levels on fish. *Trends in Ecology and Evolution, 25*, 419–427.

Snell-Rood, E. C. (2012). The effect of climate on acoustic signals: Does atmospheric sound absorption matter for bird song and bat echolocation? *The Journal of the Acoustical Society of America, 131*(2), 1650–1658.

Spencer, K. A., Buchanan, K. L., Goldsmith, A. R., & Catchpole, C. K. (2003). Song as an honest signal of developmental stress in the zebra finch (Taeniopygia guttata). *Hormones and Behavior, 44*(2), 132–139.

Stowell, D., & Sueur, J. (2020). Ecoacoustics: Acoustic sensing for biodiversity monitoring at scale. *Remote Sensing in Ecology and Conservation.* https://doi.org/10.1002/rse2.174

Sueur, J. (2002). Cicada acoustic communication: Potential sound partitioning in a multispecies community from Mexico (Hemiptera: Cicadomorpha: Cicadidae). *Biological Journal of the Linnean Society, 75*, 379–394.

Sueur, J. (2018). *Sound analysis and synthesis with R*. Springer.

Sueur, J., & Farina, A. (2015). Ecoacoustics: The ecological investigation and interpretation of environmental sound. *Biosemiotics, 8*, 493–502. https://doi.org/10.1007/s12304-015-9248-x

Sueur, J., Aubin, T., & Simonis, C. (2008). Seewave: A free modular tool for sound analysis and synthesis. *Bioacoustics, 18*, 213–226.

Sueur, J., Farina, A., Gasc, A., Pieretti, N., & Pavoine, S. (2014). Acoustic indices for biodiversity assessment and landscape investigation. *Acta Acustica United with Acustica, 100*(4), 772–781.

Towsey, M., Wimmer, J., Williamson, I., & Roe, P. (2014a). The use of acoustic indices to determine avian species richness in audio-recordings of the environment. *Ecological Informatics, 21*, 110–119.

Towsey, M., Zhang, L., Cottman-Fields, M., Wimmer, J., Zhang, J., & Roe, P. (2014b). Visualization of long-duration acoustic recordings of the environment. *Procedia Computer Science, 29*, 703–712.

Truax, B. (Ed.). (1999). *Handbook for acoustic ecology*. Cambridge Street Publishing.

Tucker, D., Gage, S. H., Williamson, I., & Fuller, S. (2014). Linking ecological condition and the soundscape in fragmented Australian forests. *Landscape Ecology, 29*(4), 745–758.

Velásquez, N. A., Moreno-Gómez, F. N., Brunetti, E., & Penna, M. (2018). The acoustic adaptation hypothesis in a widely distributed South American frog: Southernmost signals propagate better. *Scientific Reports, 8*(1), 1–12.

Wallschläger, D. (1980). Correlation of song frequency and body weight in passerine birds. *Experientia, 36*(4), 412–412.

Ward, M. P., & Schlossberg, S. (2004). Conspecific attraction and the conversation of territorial songbirds. *Conservation Biology, 18*(2), 519–525.

Wiley, R. H. (1994). Errors, exaggeration, and deception. In *Behavioral mechanisms in evolutionary ecology* (Vol. 157).

Wyman, M. T., Mooring, M. S., Mccowan, B., Penedo, M. C. T., Reby, D., & Hart, L. A. (2012). Acoustic cues to size and quality in the vocalizations of male North American bison, Bison bison. *Animal Behaviour, 84*(6), 1381–1391.

Zeyl, J. N., den Ouden, O., Köppl, C., Assink, J., Christensen-Dalsgaard, J., Patrick, S. C., & Clusella-Trullas, S. (2020). Infrasonic hearing in birds: A review of audiometry and hypothesized structure–function relationships. *Biological Reviews, 95*(4), 1036–1054.

Chapter 2
The Acoustic Complexity Indices (ACIs)

Abstract The acoustic complexity indices represent a family of metrics formulated to extract information from sonic matrices. These indices are based on the differences of acoustic energy measured along a temporal interval of a sonic matrix across all the spectral lines ($ACIf_t$) and along each individual spectral line ($ACIt_f$) obtaining respectively a temporal and a frequential sonic signature. The distribution of information in time and in frequency returns respectively an $ACIf_t$ evenness and an $ACIt_f$ evenness that, when combined in a ternary code with $ACIf_t$, compose the codes of *Ecoacoustic Events* (EE). The Ecoacoustic Events are defined as an emergent aggregation of sonic information that assumes a sense for a species or for an ecological process. In order to obtain realistic data, every sonic matrix must be cleaned from low energetic elements, mainly artifacts, by applying adapt energy filters that are also used to delimit *near* and *far acoustic fields* contributing to better understand the relationships between sounds and the environment.

A clumping procedure can aggregate original data before the application of ACI metrics. The *Sonic Signature Dissimilarity* (SSD) is used to assess the distance in $ACIt_f$ between two sonic matrices. The application of the Shannon theory returns the value of entropy of ecoacoustic events.

The application of different temporal scales to the ACI computation allows to overpass the uncertainty in the selection of the more adapt temporal resolution obtaining a fractal dimension of ecoacoustic events and of sonic dissimilarity.

Keywords Acoustic Complexity Index · Clumping · $ACIf_t$ · $ACIt_f$ · $ACIf_{t\ evenness}$ · $ACIt_{f\ evenness}$ · Ecoacoustics events · Fractal dimension

2.1 A Short History of ACI

From 2006 to 2008, one of us (AF) was the supervisor of a PhD student at Urbino University, and at that time, his interest in ecoacoustics was really at the beginning. However, his study of behavioral ecology crossed paths with the sounds broadcasted by birds. In particular, he was interested to investigate the distribution of

A. Farina, P. Li, *Methods in Ecoacoustics*, Frontiers in Ecoacoustics 1,
https://doi.org/10.1007/978-3-030-82177-7_2

sounds across a landscape and to discover the level of acoustic aggregation in space and time.

To do this, he assigned to the PhD student the task to build across an agricultural landscape a cableway of 500 m long, upgraded in the successive year at 900 m, at which to suspend a magnetic tape recorder (a Sony Dat 1100 equipped with a Telinga Pro 5 W microphone) that was moved by a winch. At that time, a cableway seemed a good strategy to deploy in short time a recorder at different positions (50/100 m distant each to the other) in the field evaluating the effect of landscape mosaic on acoustic activity of soniferous species.

Few years later, this approach was abandoned thanks to the availability on the market of field autonomous recorders with the possibility of their deployment in a lattice (e.g., Farina & Pieretti, 2014).

The acoustic survey was carried out at fixed distances when the cableway was stopped because the sound produced by wheels and cable vibration was masking the natural sounds. The study area was located close to a small river responsible for a high geophonic sound, especially after heavy rains. The idea to filter away the sound of the flowing water and to isolate the biological sounds was at the basis of a new metric that we called Acoustic Information Index (Morri, 2006) and then changed to the Acoustic Complexity Index (ACI) (Pieretti et al., 2011). Indeed, this metric measures some information conveyed by sounds and should be named *acoustic information index*, returning to the original labeling! The reason of the abandonment of the original name largely depended on the contradictory use of the term information (see for a discussion on this point, Mingers & Standing, 2018).

ACI was based on the fact that regular sounds (f.i., some continuous sounds) when analyzed at a suitable temporal scale had the same value of amplitude and that was easily filtered away by a subtraction carried by a simple *ad hoc* script. In this way, the first attempt to clean an acoustic file when transformed from a temporal domain by a Fourier transform into a frequency domain was at the basis of ACI project. The difference between amplitude values was calculated at different temporal scale and discarding all the values emerged from the difference below a certain energy threshold. The first results were very encouraging, and ACI was calculated as difference between pitches created by biophonies (mainly birds). After this first experience, ACI equation was improved by appropriate usage of time aggregation that we call "clumping" and after a suitable filter (that we call energy filter) to exclude every source of disturbing nonenvironmental signals. Today ACI metric is very popular and adopted by several scholars worldwide in terrestrial (f.i., Gasc et al., 2013; Fuller et al., 2015; Farina et al., 2018; Colonna et al., 2020) and in aquatic systems (f.i., Duarte et al., 2015; Bertucci et al., 2016, 2020; Bolgan et al., 2018; Ceraulo et al., 2018; Bittencourt et al., 2020; Davies et al., 2020). At the same time, two indices were used, the ACI along each spectral line ($ACIt_f$) and a second across all spectral lines for each temporal step ($ACIf_t$). The two indices explore the patterns present in a sonic matrix moving along spectral lines and across a temporal interval, respectively. $ACIt_f$ is utilized to describe the spectral (sonic) signature, and $ACIf_t$ describes the variation of sonic information along the time (temporal [sonic] signature).

Both the metrics assume values according to the dimension of the file, its length, and the Fourier transform utilized.

ACI operates on frequencies aggregated according to a linear scale but not in a logarithmic scale as done by traditional acoustics where amplitude is converted in decibels.

2.2 Mathematical Equations

Information, considered a polymorphic and a polysemantic concept, is used in sonic assessment for its character of convertible coin (Stonier, 1996). The ACI is based on the idea that the sonic information embedded into a sound sequence may be handled mathematically.

After the migration of sound from a temporal domain to a frequential domain using the Fourier transform (specifically, a time-frequency domain analysis based on, for example, short-term Fourier transform, or STFT), data in wav format are arranged into a sonic matrix where the *y* axis represents the spectral lines (*f*) and the *x* axis the time (*t*) (Figs. 2.1 and 2.2) and then processed by the ACI algorithms. The ACI metric is composed of two different indices: a first index that we call $\mathrm{ACIt_f}$ calculates the amount of information present along every spectral line at selected intervals of time. A second index that we call $\mathrm{ACIf_t}$ calculates the amount of information across all the spectral lines at each temporal step (we use this convention f = frequency, t = time).

The two indices are calculated according to the following equations (Pieretti et al., 2011; Farina et al., 2016, 2018):

$$\mathrm{ACIt_f} = \sum_{t=1}^{n_t-1}\left(\frac{\left|a_{t,f} - a_{t+1,f}\right|}{a_{t,f} + a_{t+1,f}}\right) \tag{2.1}$$

$$\mathrm{ACIf_t} = \sum_{f=1}^{n_f-1}\left(\frac{\left|a_{t,f} - a_{t,f+1}\right|}{a_{t,f} + a_{t,f+1}}\right) \tag{2.2}$$

where $a_{t,f}$ is the amplitude of each pulse along the spectral line f at the time step t; n_t is the total number of temporal frames that depend on the temporal aggregation parameter used (e.g., for an acoustic signal sampled at 48 kHz and with an STFT window size of 1024, for temporal aggregation 1 s, $n_t = \lfloor 48{,}000/1024 \rfloor = 46$); and n_f is the number of spectral lines (e.g., in the example, $n\_f = 512$). The aggregation is called clumping, and the value of aggregation may return emergent characters when the sonic patterns are well structured.

$\mathrm{ACIt_f}$ and $\mathrm{ACIf_t}$ are calculated according to a normalization procedure, and their values (for each confrontation) range from 0, when the values a_i and a_{i+1} are identical, to 1, when one of the two elements of confrontation is = 0. However, to avoid an edge effect, the value 1 is not accepted in the computation. See the next session devoted to the elimination of the edge effect.

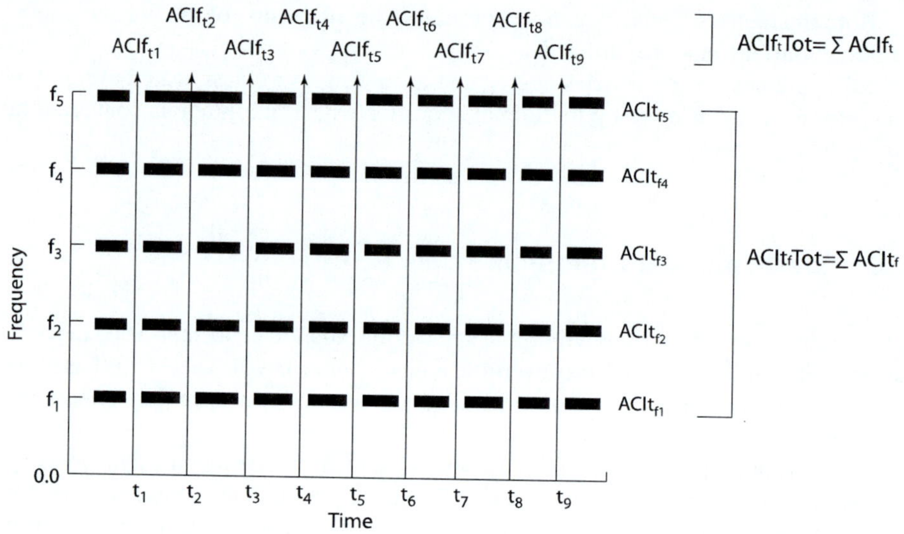

Fig. 2.1 ACI metrics are calculated along each spectral line (ACIt$_f$) and at each temporal step across all frequencies (ACIf$_t$). Here an example based on five spectral lines and nine temporal steps. The number of spectral lines depends on the window length when performing the STFT. The frequency resolution (in other words, the difference in frequency between subsequent spectral lines) depends on both the sampling frequency of the wav file under analysis and the window length when performing STFT

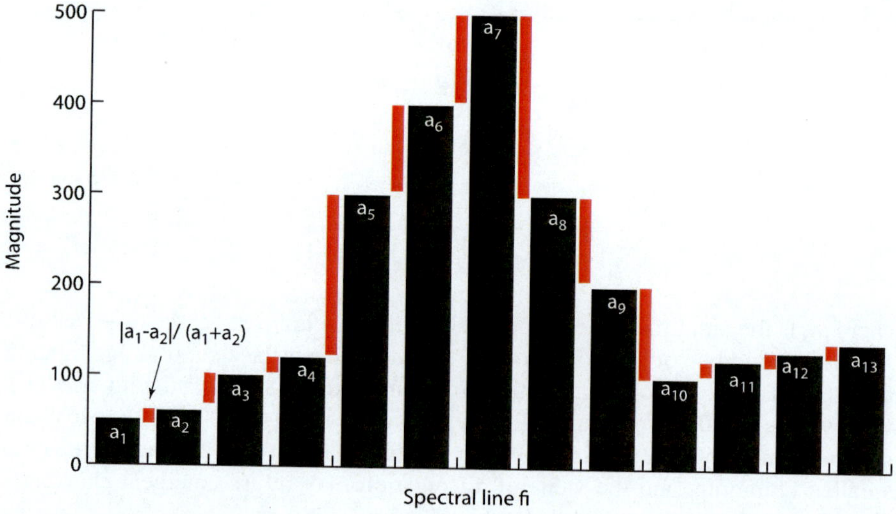

Fig. 2.2 ACIt$_f$ is the index utilized to extract the acoustic information from each spectral line *fi* calculating the normalized absolute difference between two adjacent pulses of $a_{t,f}$ amplitude for a temporal interval *t*

2.3 Avoiding the Edge Effect

ACI metrics are based on the calculation of (information) difference between successive elements of a sonic matrix. The heterogeneity present either along a single spectral line or across all frequencies at each temporal step opens a confrontation between elements that may have a sonic energy equal to 0 with elements with a sonic energy >0.

According to the ACI equations, the confrontation between an element of 0 value with a successive element with a value >0 produces an ACI = 1, that is the maximum that it is possible to obtain when this algorithm is applied. For instance, if an element (a_1) is = 0 and a second element $a_2 = 50$, we obtain $|a_1 - a_2|/(a_i + a_2)$, and using 0 and 50, we obtain $|0 - 50|/(0 + 50) = 1$.

When applicable to a matrix, we could have the maximum of ACI at the border of a signal, and this could affect the final results. After this simple and valid reason, ACI equation is applied only between two or more sonic cells that have no 0 values (Fig. 2.3). With this strategy, the sonic matrix is also cleaned of every isolated elements ≠0 erasing all the information that cannot be transformed into a meaning for a listener.

2.4 Sonic Signatures

With the term *sonic signature,* we indicate an aggregated value of $\mathrm{ACIt_f}$ or $\mathrm{ACIf_t}$ that describes the temporal or the spectral distribution of sonic information.

In particular, we define *Spectral Sonic Signature* (SSS) as the distribution of frequencies along a specific discrete interval of time sufficient to eventually encompass a sonic phenomenon. An *SSS* is obtained by aggregating the amount of information of a sonic matrix of each spectral line according to the equation:

$$\mathrm{SSS} = \sum_{1}^{n} \mathrm{ACIt_f}$$

where *n* is the number of spectral lines.

We define *Temporal Sonic Signature* (*TSS*) as the distribution of the acoustic information $\mathrm{ACIf_t}$ along a temporal interval according to the equation:

$$\mathrm{TSS} = \sum_{1}^{t} \mathrm{ACIf_t}$$

where *t* is an interval of time.

According to the sonic phenomenon of interest, we distinguish *Bioacoustic* (*SSSb, TSSb*), *Geoacoustic* (*SSSg, TSSg*), *Anthropoacoustic* (*SSSa, TSSa*), *Technoacoustic* (*SSSt, TSSt*), and *Ecoacoustic signature* (*SSSe, TSSe*). These categories refer to typical call, song, contact call, vocalization of species, or the isolated

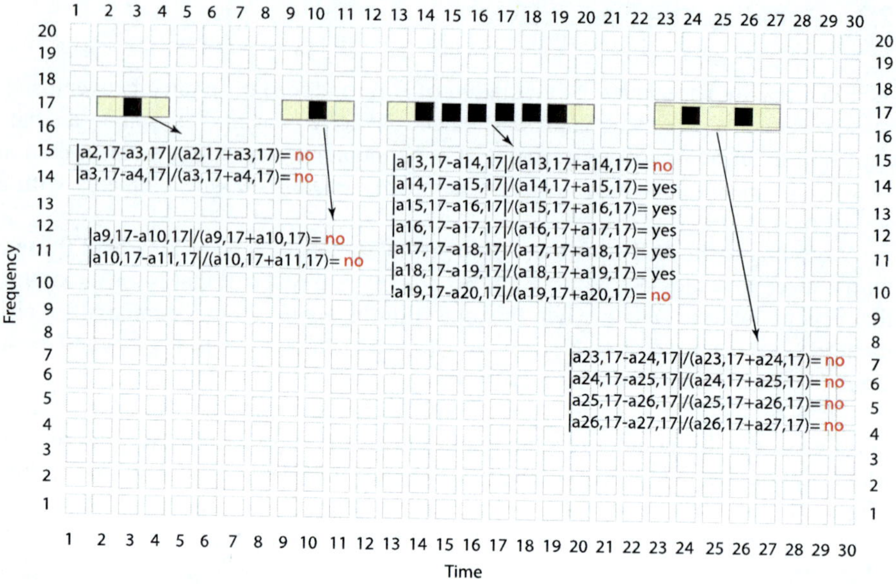

Fig. 2.3 Example on the way $ACIt_f$ operates in presence of cells with a value of amplitude = 0. In this case, we have considered the spectral line 17, and along this line, we have applied $ACIt_f$ metric. Every confrontation in which one of the two cells is equal to 0 is not considered in the calculus of $ACIt_f$, totaling in this case only five confrontations. This is applied in the same way to the calculus of $ACIf_t$

sound of atmospheric events like thunders, water falls, vehicle engine, human voice, or frictions between matter in movement.

In particular, *Ecoacoustic signature* refers to a portion of a sonic matrix that describes an acoustic community or a complex phenomenon like an animal chorus or a sonotope.

The *Ecoacoustic signature* is the footprint of an ecoacoustic event and describes the distribution in the frequential domain of aggregated acoustic objects that have an intrinsic ecological relevance. For instance, a dawn chorus composed of different species songs, a sequence of alarm calls in presence of a predator, or a portion of a sonoscape in which are present sounds of different origin. We present two examples: the sound of a thunderstorm and the sound of a dawn chorus of singing birds.

The sound of a thunderstorm is a geophonic event composed of a sequence of individual thunders that in turn have a specific geoacoustic signature (Fig. 2.4).

A dawn chorus is a biophonic event composed by different bioacoustic signatures each representing a soniferous species (Fig. 2.5) or by an aggregation of the same sonic signature of a species (f.i., the chorus of starlings (*Sturnus vulgaris*) at a dusk roosting).

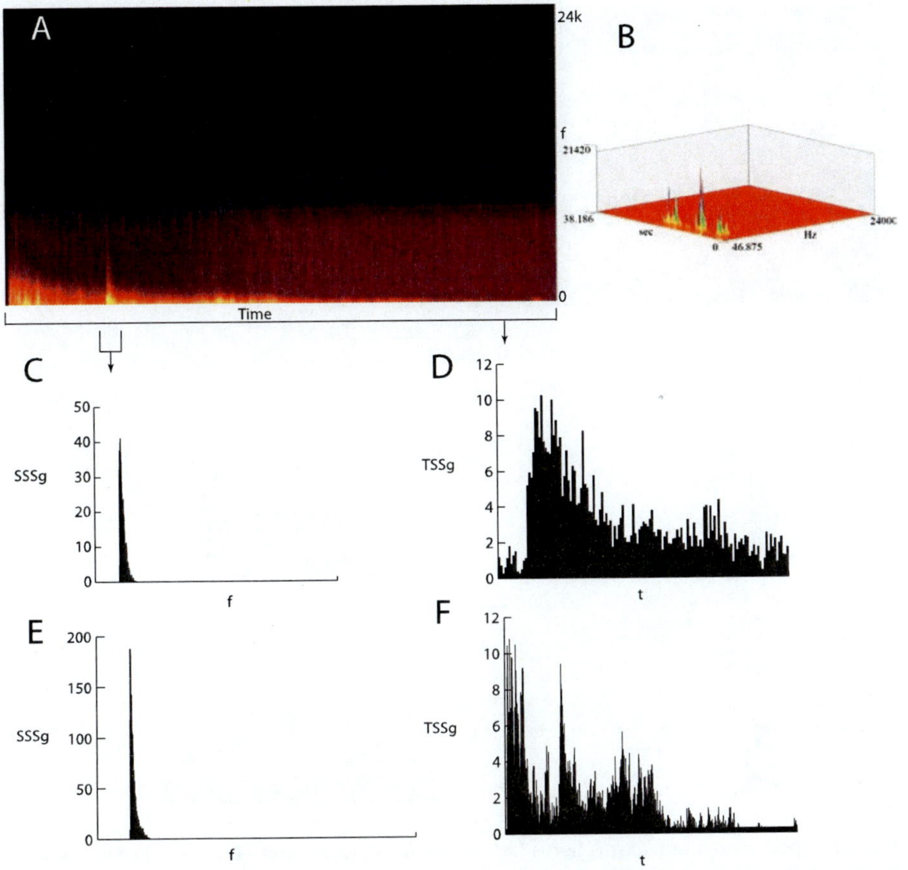

Fig. 2.4 Spectrogram of a thunderstorm (**A**) and its 3D representation (**B**). Spectral *SSSg* (**C**) and temporal TSSg (**D**) sonic signature of a thunder, and spectral *SSSg* (**E**) and temporal TSSg (**F**) sonic signature of the sequence of thunderstorm (Moncigoli, Fivizzano, It, 5/12/2020, 16:36, 44°12′ N, 10°05′ E)

2.5 Clumping

Clumping is a procedure of temporal aggregation of data before the application of the ACI metrics. This procedure has been assessed to search for aggregation patterns that are hidden when a fine-resolution Fourier transform is applied. In particular, clumping assumes importance in searching for the best temporal resolution at which to consider the elements of the sonic matrix and can be used to discover the level of complexity (by using a fractal dimension) hidden away into a sonic matrix.

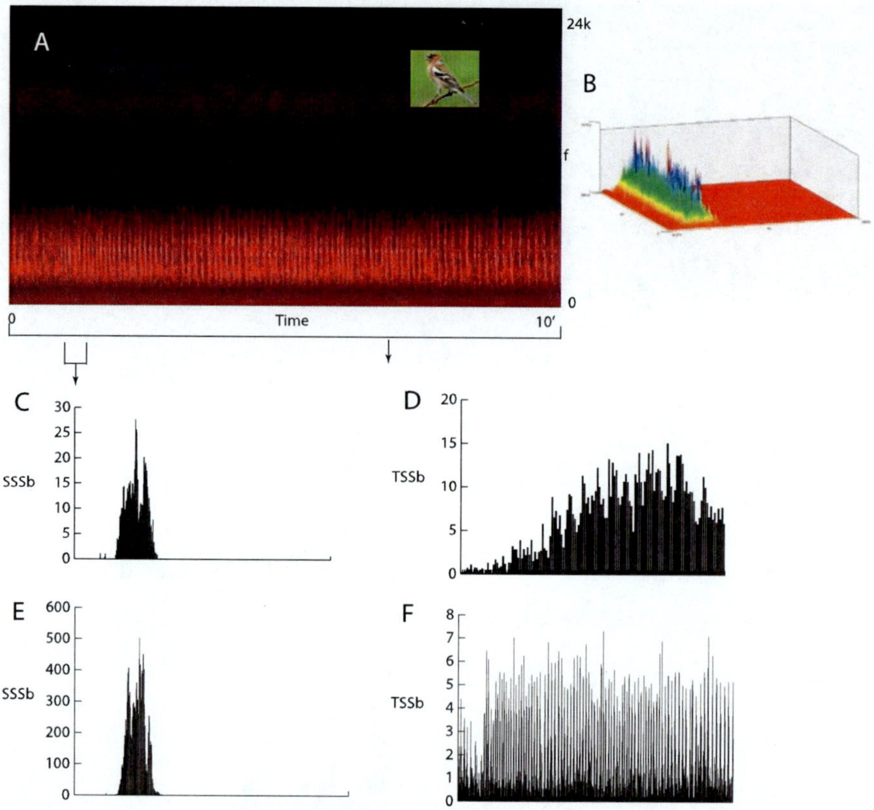

Fig. 2.5 Spectrogram of a dawn chorus (**A**) and its 3D representation (**B**). Spectral *SSSb* (**C**) and temporal *TSSb* (**D**) sonic signature of a chaffinch song, and spectral *SSSb* (**E**) and temporal *TSSb* (**F**) sonic signature of the sequence of songs of the same species in an interval of 10 min (Lama, Sasso Fratino Regional Park, 16 May 2017, 03:30 CEST, 43°11′ N, 11° 47′ E) (Courtesy of Gianni Pavan)

With the clumping procedure, we modify the resolution of the sonic matrix obtained from a Fourier transformation. The sampling design of a digital recorder is created to have the max resolution of the sonic data, but we have no evidence that the adopted resolution (f.i., 96 kHz, 48 kHz, 32 kHz, etc.) is the optimum for the research goal especially if our interest goes beyond bioacoustic competencies.

In Fig. 2.6 is illustrated the procedure of clumping computation.

The selection of the clumping resolution depends on the goal of the research and on the characters of the sonic matrix. The application of a sequence of clumping values allows to discover the internal complexity of a sonic matrix, and different clumping values return different $ACIt_f$ and $ACIf_t$ values as reported in the examples from Figs. 2.7, 2.8, 2.9, 2.10, 2.11, 2.12, and 2.13.

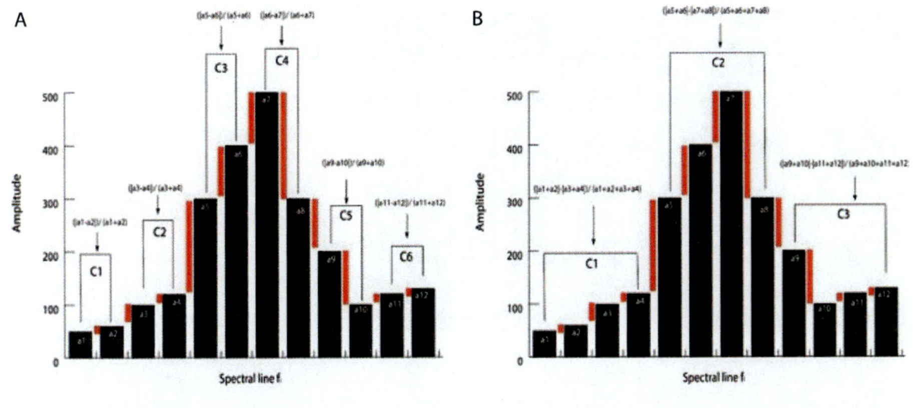

Fig. 2.6 Data may be analyzed using an aggregation or clumping. Clumping may be utilized to discover the hidden structure of a sonic matrix. The size of the clumping depends on the goal of investigation. In **A** when no clumping is utilized (default) and in **B** when applied a value 2 of clumping. In **A** the confrontation has been made between two elements, in **B** the confrontationhas been obtained between four elements at time

Every level of clumping produces a different sonic signature, although these differences are not always significantly different and reveal a different content of released information.

In the example described in Fig. 2.7, the clumping has been applied to a dawn chorus of a temperate montane forest. Starting with the clumping of a single point until the clumping of 20 elements at time, it is possible to observe a rising value of $ACIf_t\_Tot$ ($ACIf_t\ Tot = \Sigma ACIf_t$) until cluster of 11–12, then a gradual decline. The sonic signature shows a gradual change in the shape starting from clump 1 to 16 onward. Changing the value of clumping, the sonic signature remains similar only inside a restrict range and then is deeply modified.

Figure 2.8 shows an insect chorus from Madagascar. The fourth category of clumping seems the most informative.

In the example of Fig. 2.9 referring to a morning singing activity in temperate shrubland, $ACIt_f\_Tot$ increases until the value of clumping of 70 demonstrating great irregularities at all the scale.

In Fig. 2.10, the information during a rainy period decreases a lot after the first clumping class returning the evidence of the uniformity of the process with the increase of the clumping category.

Figure 2.11 represents a sequence of chaffinch songs where a sinusoidal-like curve for the first clumping categories emerges.

In Fig. 2.12, a wind sequence shows an increase moving from clumping 1 to clumping 2 and then a slow decrease until clumping 10.

In Fig. 2.13, a sonic matrix of insects from Madagascar shows a quick increase of $ACIt_f$ until clumping 4 and then a slow decrease. In this figure, a sonic matrix of insect with a great regularities in the acoustic signature shows at the change of clumping a behavior similar to a windy day.

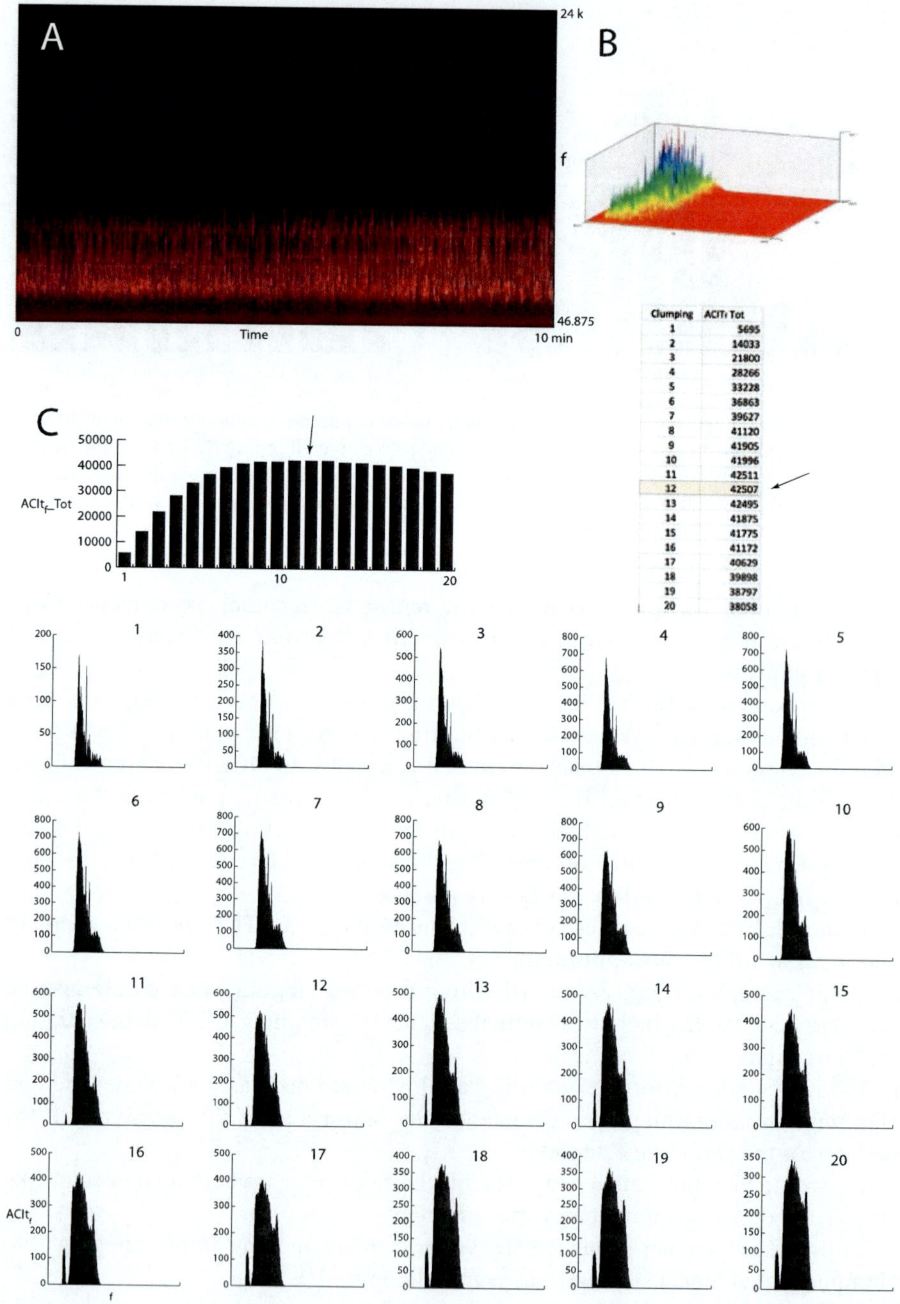

Clumping	ACITf Tot
1	5695
2	14033
3	21800
4	28266
5	33228
6	36863
7	39627
8	41120
9	41905
10	41996
11	42511
12	42507
13	42495
14	41875
15	41775
16	41172
17	40629
18	39898
19	38797
20	38058

Fig. 2.7 Trend of $ACIt_f$ after the application of the clumping procedure for values from 1 to 20 in a Sasso Fratino dawn chorus on May 11, 2017, 05:30 a.m. (43°11′ N, 11° 47′ E) (Courtesy of Gianni Pavan)

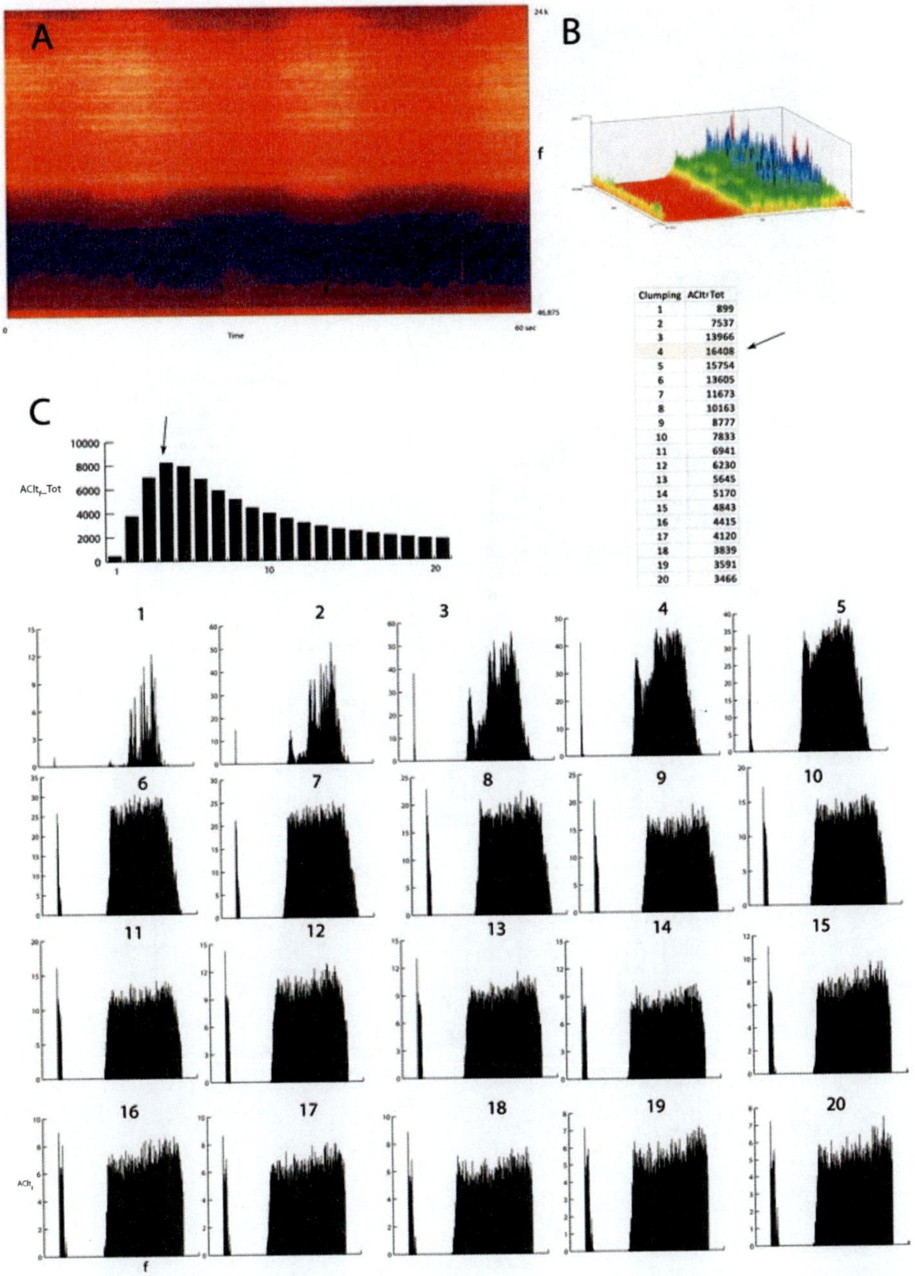

Fig. 2.8 Insect chorus in a Madagascar plot (Courtesy of Anne Axel). The highest value of $ACIt_f$ has been obtained at a clumping value of 4

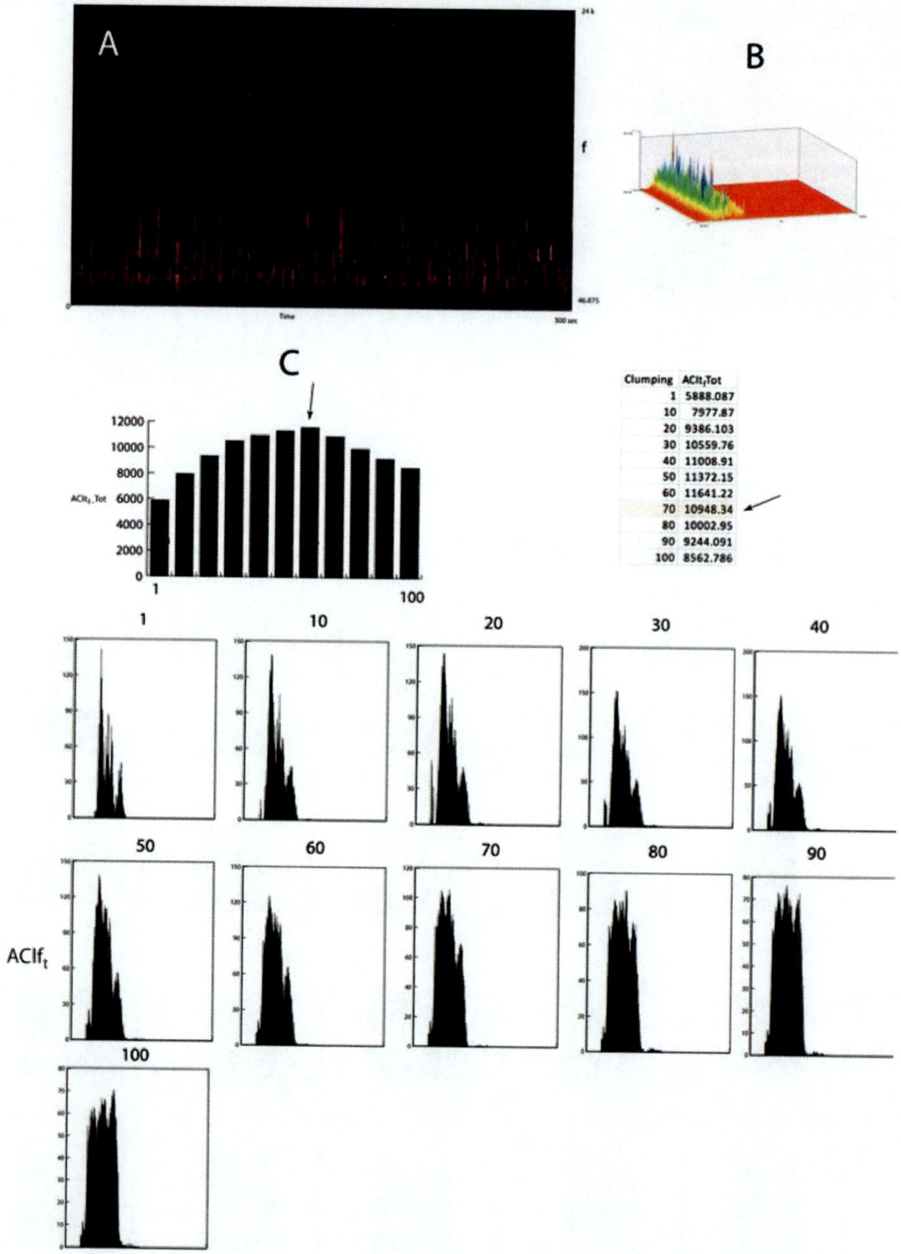

Fig. 2.9 Trend of $ACIt_f$ after the application of the clumping procedure for values from 1 to 100 at Agnolo lake location (43°13′N, 10°04′E) in morning on May 23, 2020, 09:00 a.m. The highest level of information is achieved at clumping 70

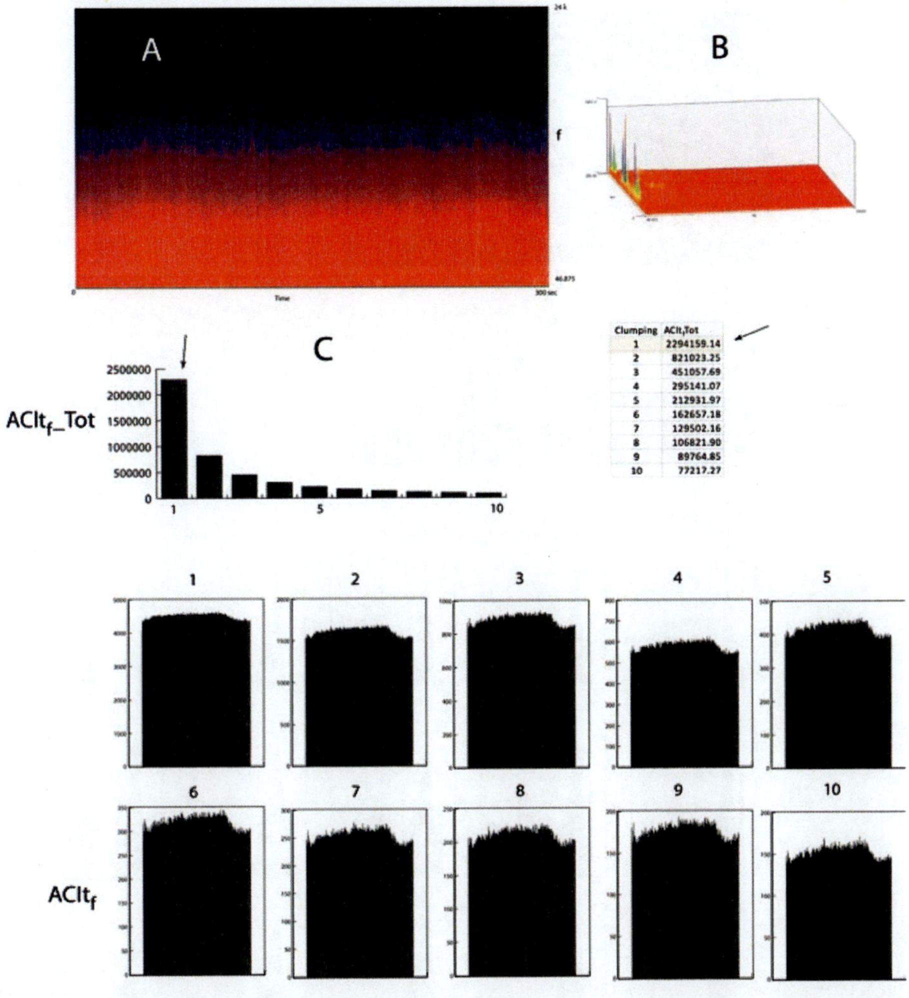

Clumping	ACIt,Tot
1	2294159.14
2	821023.25
3	451057.69
4	295141.07
5	212931.97
6	162657.18
7	129502.16
8	106821.90
9	89764.85
10	77217.27

Fig. 2.10 Trend of $ACIt_f$ after the application of the clumping procedure for values from 1 to 10 in Ospedale (Fivizzano) wood (44°14′, 10°07E) in a rainy day on December 9, 2020, 04:01 a.m. The highest level of information is achieved at clumping 1

Clumping procedure results useful when we would like to find the best temporal resolution at which we obtain the maximum of sonic information. At the same time, the analysis of the curve obtained plotting $ACIt_f$ values against the clumping value expresses evident patterns according the internal structure of a spectrogram. In the case of great regularity, the curve increases very quickly in few clumping steps and then decreasing more gently. In cases of great internal heterogeneity, the curve has a similar trend around the maximum.

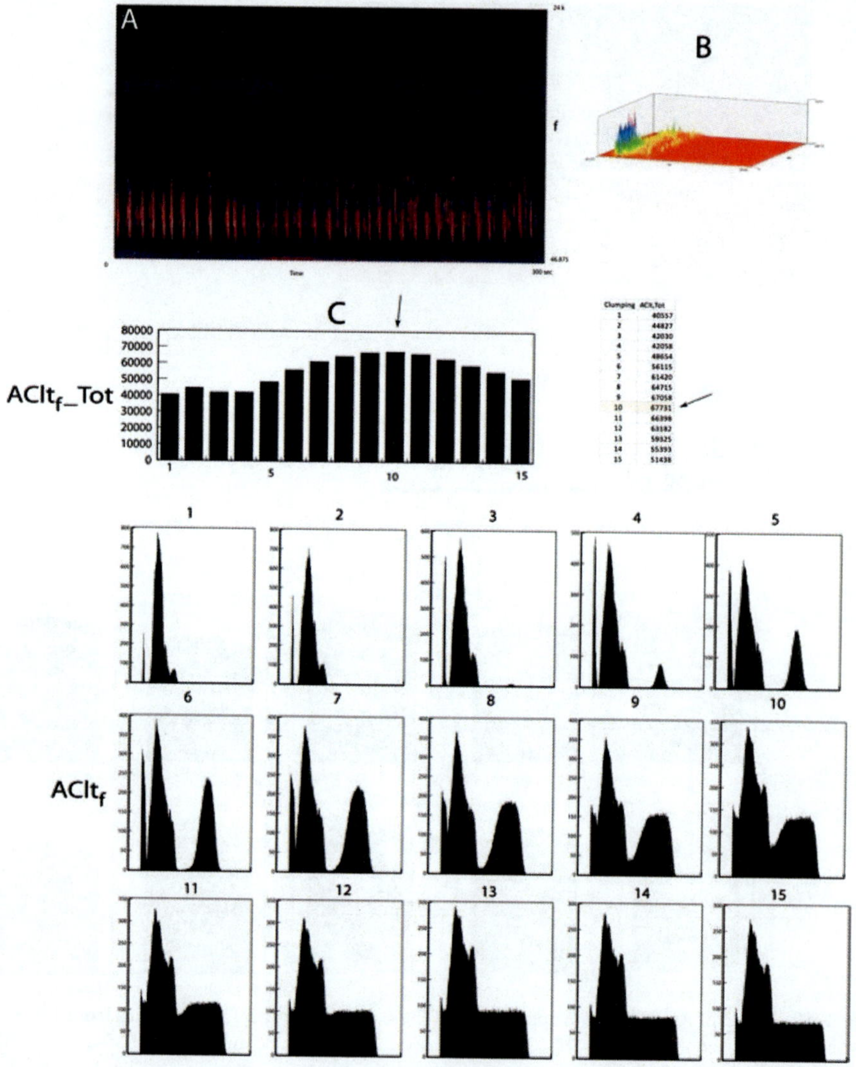

Fig. 2.11 Trend of $ACIt_f$ after the application of the clumping procedure for values from 1 to 15 in Sasso Fratino location (44°13′ N, 11°47′ E) (Courtesy of Gianni Pavan) during a Chaffinch (*Fringilla coelebs*) song performance on June 1, 2017, 05:30 p.m. The highest level of information is achieved at clumping 10

Increasing the clumping class, the total $ACIt_f$ decreases (from 1 to 0), but in some cases, this is not true especially when the internal heterogeneity of the acoustic matrix is distributed in a not regular way.

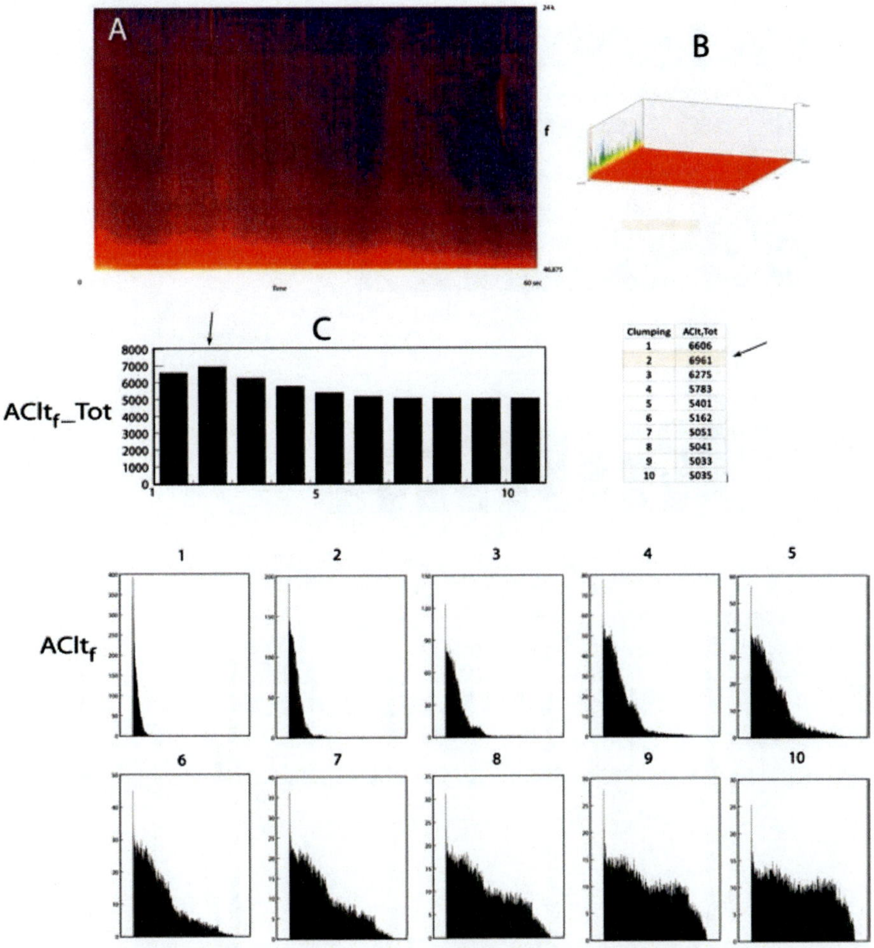

Fig. 2.12 Trend of $ACIt_f$ after the application of the clumping procedure for values from 1 to 10 in a Madagascar forest windy day on January 7, 2013, 03:00 p.m. (Courtesy of Anne Axel). The highest level of information is achieved at clumping 2

According to the examples, it emerges that the application of clumping at different resolution to the FFT matrix returns values according to the internal structure of the matrix. In synthesis, when we have a regular sonic matrix like the one created by wind or rain, we expect that the model is fully respected. But, when the internal structure is highly variable, we can observe an oscillation of the $ACIt_f$_Tot. In the

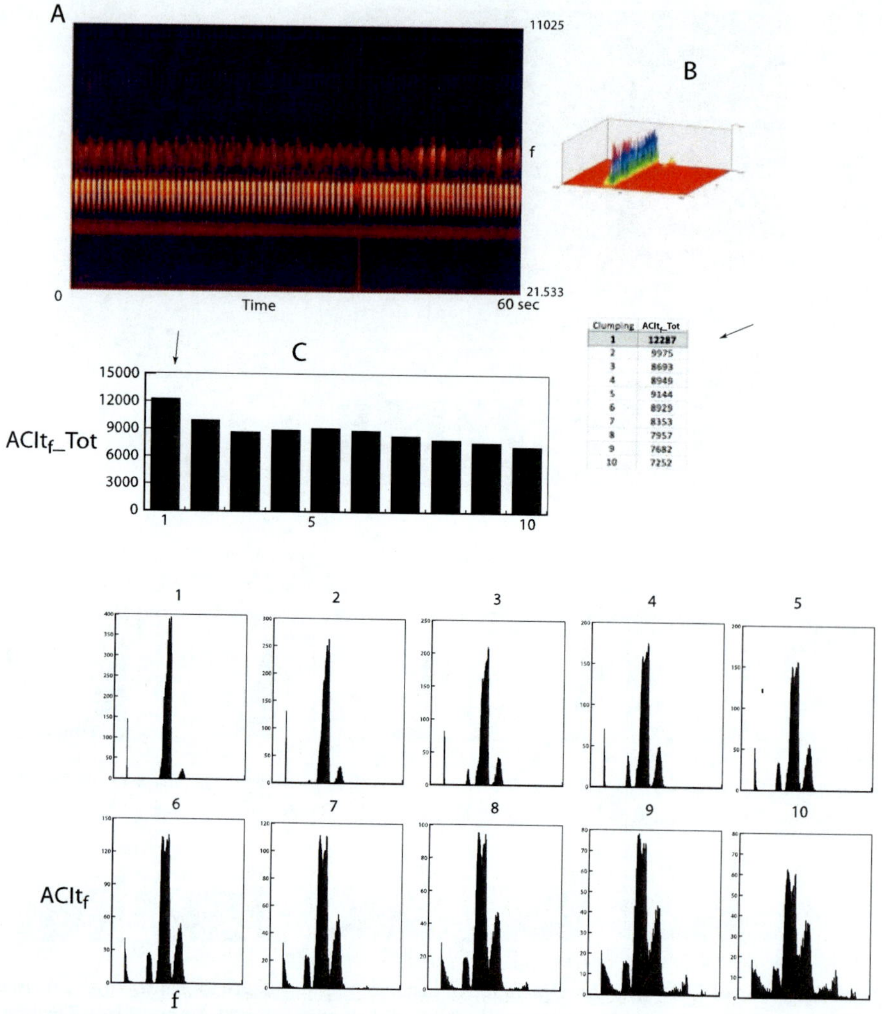

Clumping	$ACIt_f$_Tot
1	**12287**
2	9975
3	8693
4	8949
5	9144
6	8929
7	8353
8	7957
9	7682
10	7252

Fig. 2.13 Trend of $ACIt_f$ after the application of the clumping procedure for values from 1 to 10 in a Madagascar forest on January 8, 2013, 03:00 a.m. (Courtesy of Anne Axel), with a prevalent insect sounds. The highest level of information is achieved at clumping 1

case of a dense chorus as the one reported in Fig. 2.6, clumping values increase until the twelfth clump. This is an evidence that the sonic matrix has a very complex organization.

Scaling clumping allows to identify the internal structure of a spectrogram. This could be considered as a new procedure to be applied in the ecoacoustic monitoring.

2.6 $ACIt_f$ Evenness and $ACIf_t$ Evenness

The distribution of $ACIt_f$ and $ACIf_t$ respectively along each spectral line and across all the frequencies is obtained adopting the Levins evenness index B (Levins, 1968; Hurlbert, 1978; Farina et al., 2016, 2018).

$$B_{ACIt_f} = \frac{1}{\sum_{f=1}^{n_f} p_f^2} \tag{2.3}$$

$$B_{ACIf_t} = \frac{1}{\sum_{t=1}^{n_t} p_t^2} \tag{2.4}$$

where p_f is the relative importance of $ACIt_f$ along the fth spectral line (i.e., $p_f = ACIt_f / \sum_{f=1}^{n_f} ACIt_f$), and p_t is the relative importance of $ACIf_t$ along the tth temporal frame (i.e., $p_t = ACIf_t / \sum_{t=1}^{n_t} ACIf_t$) (Figs. 2.14 and 2.15). The standardized measure is:

$$ACIt_{f\,evenness} = \frac{B_{ACIt_f} - 1}{n_f - 1} \tag{2.5}$$

$$ACIf_{t\,evenness} = \frac{B_{ACIf_t} - 1}{n_t - 1} \tag{2.6}$$

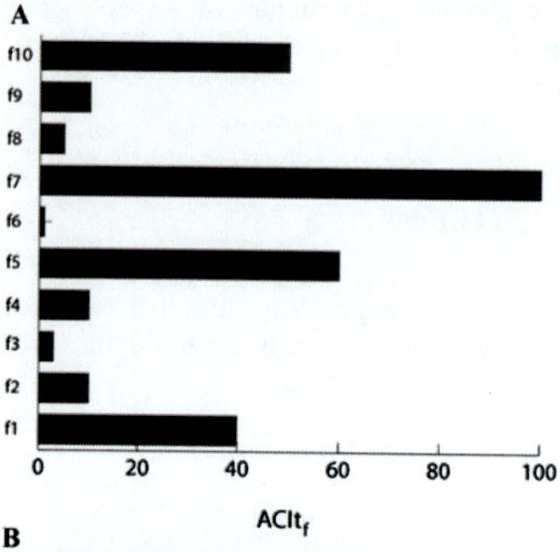

	ACIt$_f$	pi	pi*F
	50	0.17301038	0.0299:259
	10	0.03460208	0.001:973
	5	0.01730104	0.0002'933
	100	0.34602076	0.1197:037
	1	0.00346021	1.1973:-05
	60	0.20761246	0.0431'293
	10	0.03460208	0.001:973
	3	0.01038062	0.0001'776
	10	0.03460208	0.001:973
	40	0.1384083	0.0191:686
Tot.	289	1	4.6310:073

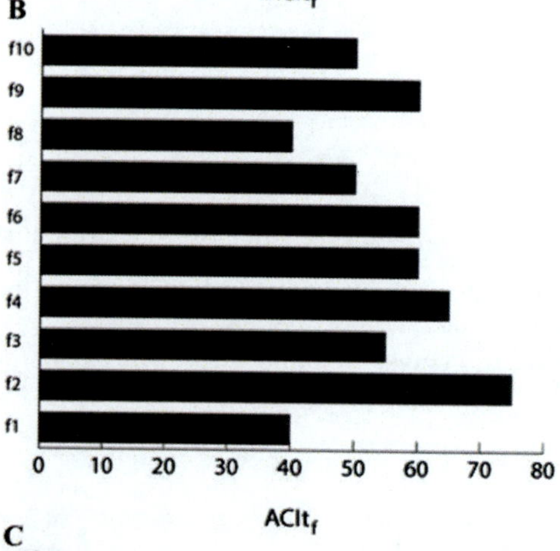

ACIt$_f$	pi	pi*pi	
50	0.09009009	0.00811622	
60	0.10810811	0.01168736	
40	0.07207207	0.00519438	
50	0.09009009	0.00811622	
60	0.10810811	0.01168736	
60	0.10810811	0.01168736	
65	0.11711712	0.01371642	
55	0.0990991	0.00982063	
75	0.13513514	0.0182615	
40	0.07207207	0.00519438	\Citf Evennes:
555	1	9.66352941	0.96

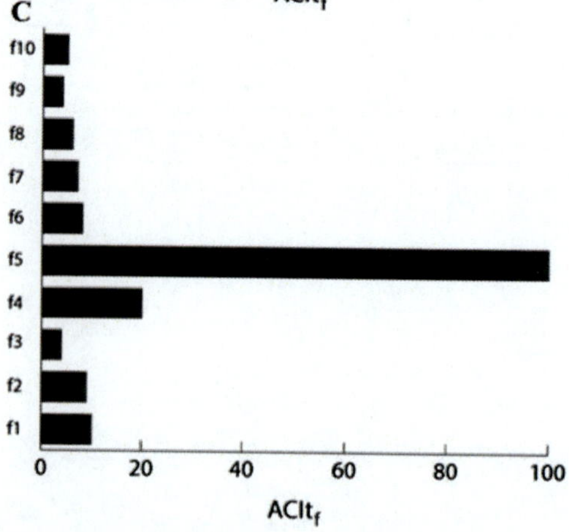

	ACIt$_f$	pi	pi*pi
	5	0.02890173	0.00083531
	4	0.02312139	0.0005346
	6	0.03468208	0.00120285
	7	0.04046243	0.00163721
	8	0.04624277	0.00213839
	100	0.57803468	0.33412409
	20	0.11560694	0.01336496
	4	0.02312139	0.0005346
	9	0.05202312	0.00270641
	10	0.05780347	0.00334:24
Tot.	173	1	2.77454343

Fig. 2.14 (A) Example of calculation of the evenness of an hypothetical heterogeneous distribution of ACIt$_f$. (B) Example of calculation of the evenness of an hypothetical uniform distribution of ACIt$_f$. (C) When the sonic signature is very asymmetric, the ACIt$_f$ evenness reaches the minimum. The examples are based on 10 hypothetical spectral lines

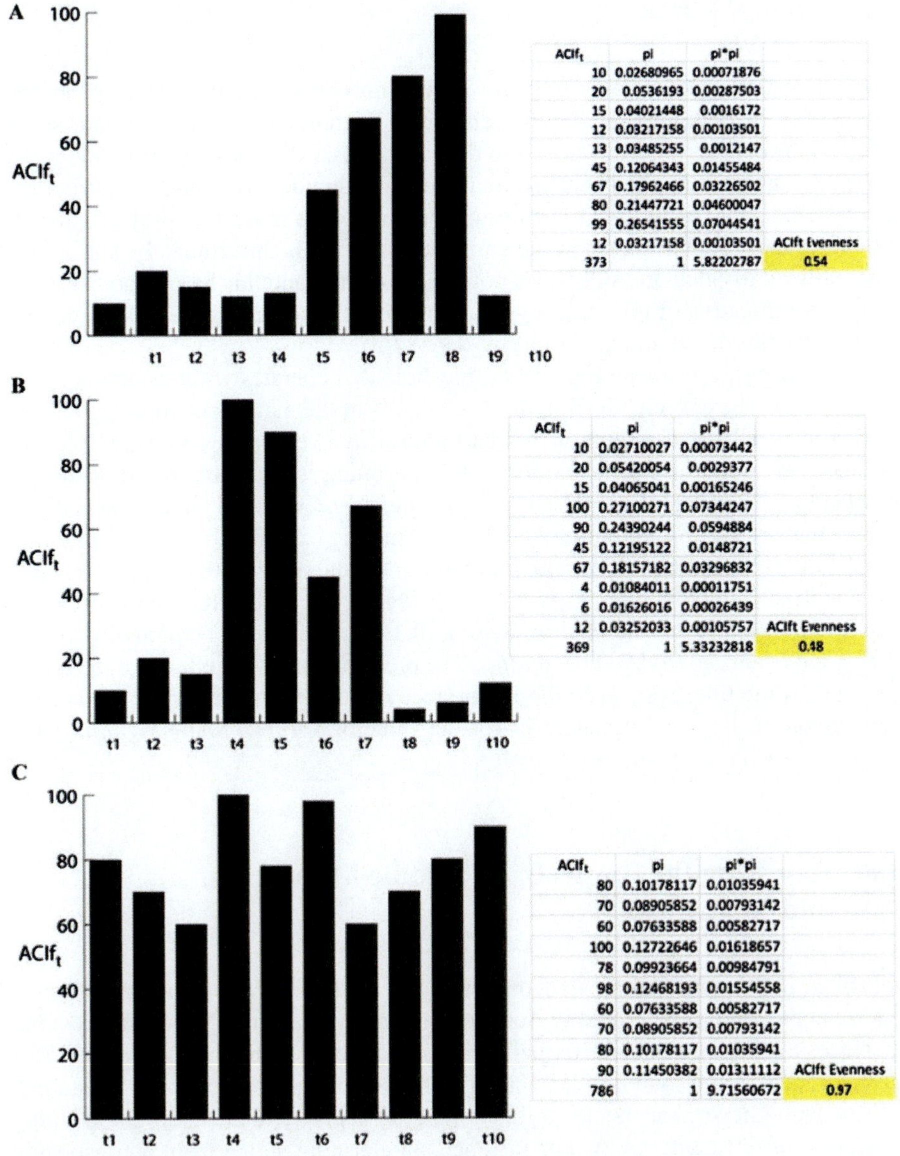

A

ACIf$_t$	pi	pi*pi	
10	0.02680965	0.00071876	
20	0.0536193	0.00287503	
15	0.04021448	0.0016172	
12	0.03217158	0.00103501	
13	0.03485255	0.0012147	
45	0.12064343	0.01455484	
67	0.17962466	0.03226502	
80	0.21447721	0.04600047	
99	0.26541555	0.07044541	
12	0.03217158	0.00103501	ACIft Evenness
373	1	5.82202787	0.54

B

ACIf$_t$	pi	pi*pi	
10	0.02710027	0.00073442	
20	0.05420054	0.0029377	
15	0.04065041	0.00165246	
100	0.27100271	0.07344247	
90	0.24390244	0.0594884	
45	0.12195122	0.0148721	
67	0.18157182	0.03296832	
4	0.01084011	0.00011751	
6	0.01626016	0.00026439	
12	0.03252033	0.00105757	ACIft Evenness
369	1	5.33232818	0.48

C

ACIf$_t$	pi	pi*pi	
80	0.10178117	0.01035941	
70	0.08905852	0.00793142	
60	0.07633588	0.00582717	
100	0.12722646	0.01618657	
78	0.09923664	0.00984791	
98	0.12468193	0.01554558	
60	0.07633588	0.00582717	
70	0.08905852	0.00793142	
80	0.10178117	0.01035941	
90	0.11450382	0.01311112	ACIft Evenness
786	1	9.71560672	0.97

Fig. 2.15 (**A**) Representation of an intermediate value of ACIf$_t$ evenness. (**B**) Representation of an intermediate value of ACIf$_t$ evenness (**C**) Representation of a high value of ACIf$_t$ evenness calculated among hypothetical 10 temporal steps. The examples are based on 10 hypothetical spectral lines

2.7 Energy Filter

A matrix often contains some levels of signals introduced by many causes like the low quality of microphones, the presence of a distant and degraded sound, some errors in the conversion from analog to digital signals, etc. This may false the results when ACI metric is applied because small but very abundant signals in a matrix may increase the value of ACI that is a normalized index. To avoid this problem, a specific filter must be introduced in the computation of ACI. Unfortunately, some software utilized to calculate $ACIt_f$ does not use such filter, and this has negative effects on the identification of the sonic signature because the sonic signature is falsed and its peculiarities do not emerge entirely.

SonoScape software applies a filter threshold that operates without subtracting a fixed value to the entire collection of the elements of the matrix but applying an "*if*" condition to all the elements of a sonic matrix that receive a confrontation. For instance, according to a Boolean logic, the reasoning is based on this sequence of instructions: "*if element* $_{i,l}$ *<=Filter value then element*$_{i,l}$*=0*". All the elements of the matrix that satisfy this condition become = 0 and are excluded by the computation of ACI metrics. The elements with an amplitude value greater than the filter value are not affected by this filter. Figure 2.16 shows the effects of filter application to a wav file dominated by the territorial song of European wren (*Troglodytes troglodytes*) and Blackcap (*Sylvia atricapilla*). The presentation of the effects of the application of some filters points to the importance of such filters either for the correct description of the sonic signature or for the value of *Spectral Sonic Signature* and *Temporal Sonic Signature*, respectively.

2.8 Assessing the Range from a Sonic Matrix: Far and Near Field

It is often important to eliminate from a wav file all the feeble sounds captured by a microphone, and this probably reduces the spatial acoustic competences but increases the possibility to better pair sounds and the environmental characters (for instance, comparing sounds to land cover using a Geographic Information Systems).

It is well known that species utilize a ranging capacity to attributing a meaning to signals of different energy. For instance, an alarming signal from a conspecific may represent a truthful signal of danger or of territorial intrusion when the signaler is close to the listener. Based on the fact that the intensity of an acoustic signal largely depends on the distance at which this signal is emitted, ranging is a process that distinguishes a near field from a far field (Farina, 2014: 72–82). In theory, it is possible to fix different distances based on a perceived level of intensity of a sound of known amplitude delimiting in such a way different more fields (Fig. 2.17). This strategy is utilized by ornithologists during bird censuses estimating the distance at which birds are heard in order to calculate the density per unit of land surveyed (Scott et al., 1981).

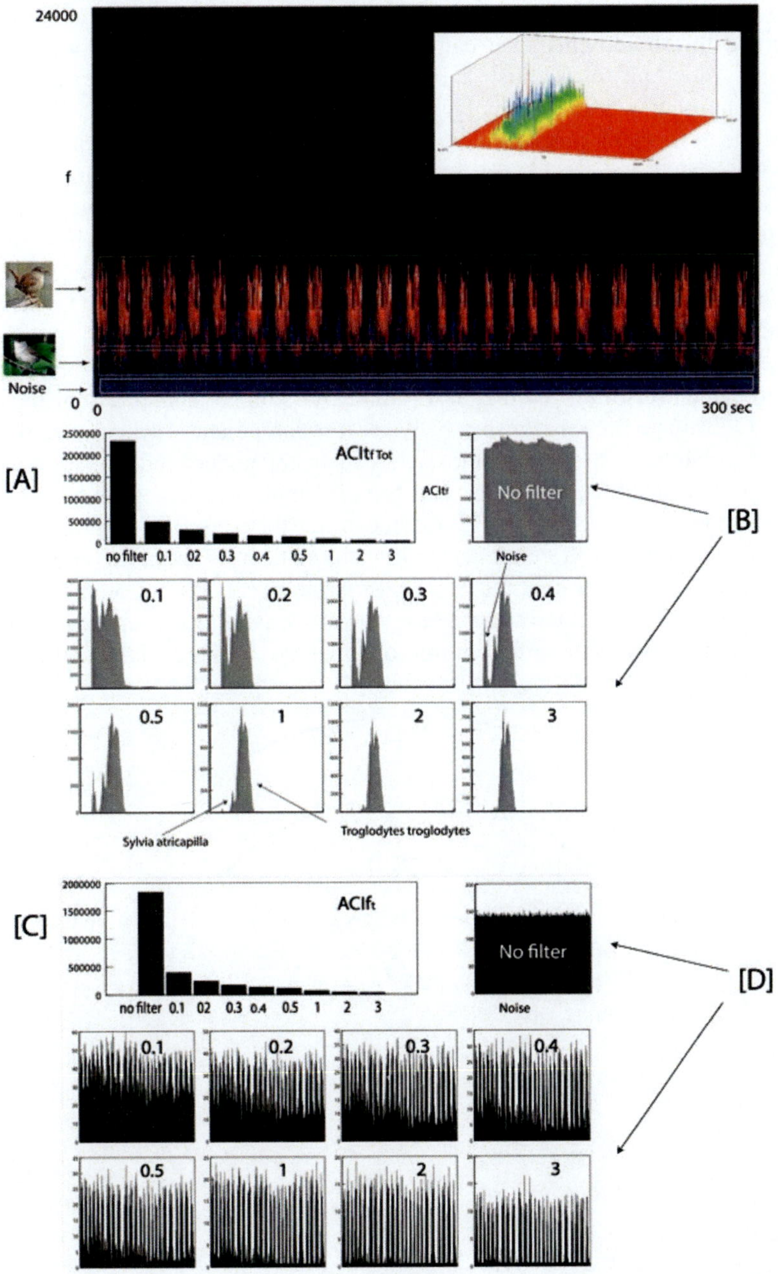

Fig. 2.16 Song sequence of European wren (*Troglodyte troglodytes*) and of blackcap (*Sylvia atricapilla*) from a beech forest (Sasso Fratino, 2018, 44°13′ N, 11°47′ E) (Courtesy of Gianni Pavan). At low frequency, a background noise emerges. In [*A*], the $ACIt_f\_Tot$ (=$\Sigma ACIt_f$) obtained applying different energy filters values and [*B*] the Spectral Sonic Signature. In [*C*], the $ACIf_t$ obtained applying different energy filters. In [*D*], the Temporal Sonic Signature

However, to delimit the acoustic field, it is necessary to empirically fix the sonic amplitude threshold. This idea can be applied in conservation actions when it becomes important to establish a border between areas at different bioacoustic diversity in relation with the typology of land cover.

The near–far-field procedure is based on the hypothesis that biophonies of related species like songbirds are uttered with a similar amplitude. Although this is a considerable approximation, it works for some groups of vertebrates like songbirds, especially in temperate and boreal biomes where many species have on average a similar sound amplitude independently to body size or residual body mass (Brumm, 2009). This fact allows to assume that the stronger the signal is, the closer the emitter is to the recording station and vice versa.

The near–far-field approach allows to delimit areas of equal sonic energy around a microphone. This may be utilized for an analysis on the abundance of focal species or to evaluate the relationship between a sonotope and the land-mosaic. With the manipulation of the sonic fields, we can better understand the structure and dynamic of acoustic communities and the distribution across the land mosaic of their component species. Figure 2.18 demostrates different ranges of microphones obtained by applying different energy filter to the data. Without the application of an energy filter (A) is evident the overlap of the sonic signals between the different microphones. After an intermediate case (B), in C, it results a separation of sonic competencies between the different microphones by applying a high energy filter.

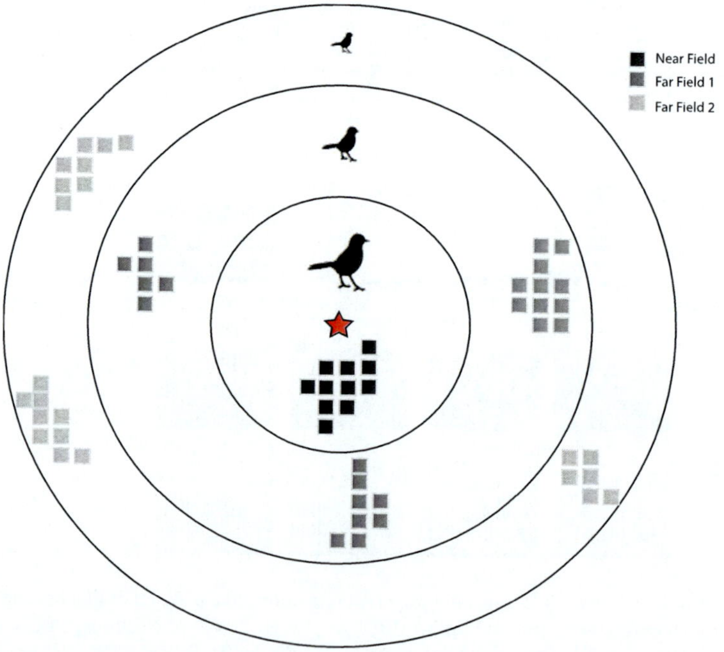

Fig. 2.17 Around a listener or a microphone is possible to delimit different acoustic spatial competencies or fields (in this case, a near field and two far field categories) based on the perceived amplitude of a signal

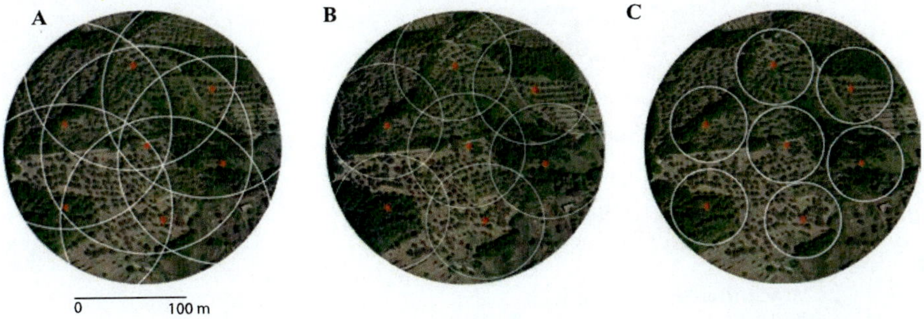

Fig. 2.18 In this simulation, an array of microphones is located at a distance of 100 m from each other. If we maintain a standard gain of microphones, we obtain an overlap between microphone ranges (**A**). If we apply a near-field filter, we can obtain a more precise resolution with the land mosaic (**B**). And finally, if we apply a severe filter (**C**), the microphone "sonic spatial competency" allows to collect information close to each microphone without overlap with the others

Figure 2.19 shows an example of application of the near- and far field to a sequence of songs of a group of chaffinches (*Fringilla coelebs*) during a time interval of 10 min. We have applied a sequence of amplitude filter of 0.05, 0.1, 0.5, 1, 2.5, 5, and 10 for the near field and of an interval of amplitude threshold from 0.01 to 0.05, 0.01 to 0.1, 0.01 to 0.5, 0.01 to 1, 0.01 to 2.5, 0.01 to 5, and 0.01 to 10 for the far field.

$ACIf_t$ in the different classes of amplitude shows a clear isolation of the levels of amplitude >1. This emerges from the confrontation with the 3D spectrogram. The songs at higher amplitude are concentrated in the first 1/3 of the 10 min spectrogram. At an amplitude of 10, only the first part of spectrogram remains. The selection of different categories of far field shows a more homogeneous distribution of $ACIf_t$ along the entire interval of time.

The near field for $ACIt_f$ shows a characteristic sonic signature that changes with the increase of the filter threshold. In the last class of amplitude (10), only a small portion of the chaffinch sonic signature persists. When $ACIt_f$ is applied without filters (0), the sonic signature appears homogeneous along all the classes.

The technique to separate signals at different amplitude can find several applications in the study of sonoscapes. In particular, with this procedure is possible to delimit areas homogeneous for levels of sonic amplitude, and this creates the conditions to delimit more precisely the sonotopes. Other possibilities of applications are are represented in the study of phenological development of acoustic activity along a season. The total amount of a near field and of a far field can be an indicator of the character of each sonotope and can explain the distribution of soniferous species in a landscape and their dynamics. For instance, with the utilization of a matrix of microphones, it is possible to create a spatial representation of the different sonotopes and to distinguish the areas at different density of singing species.

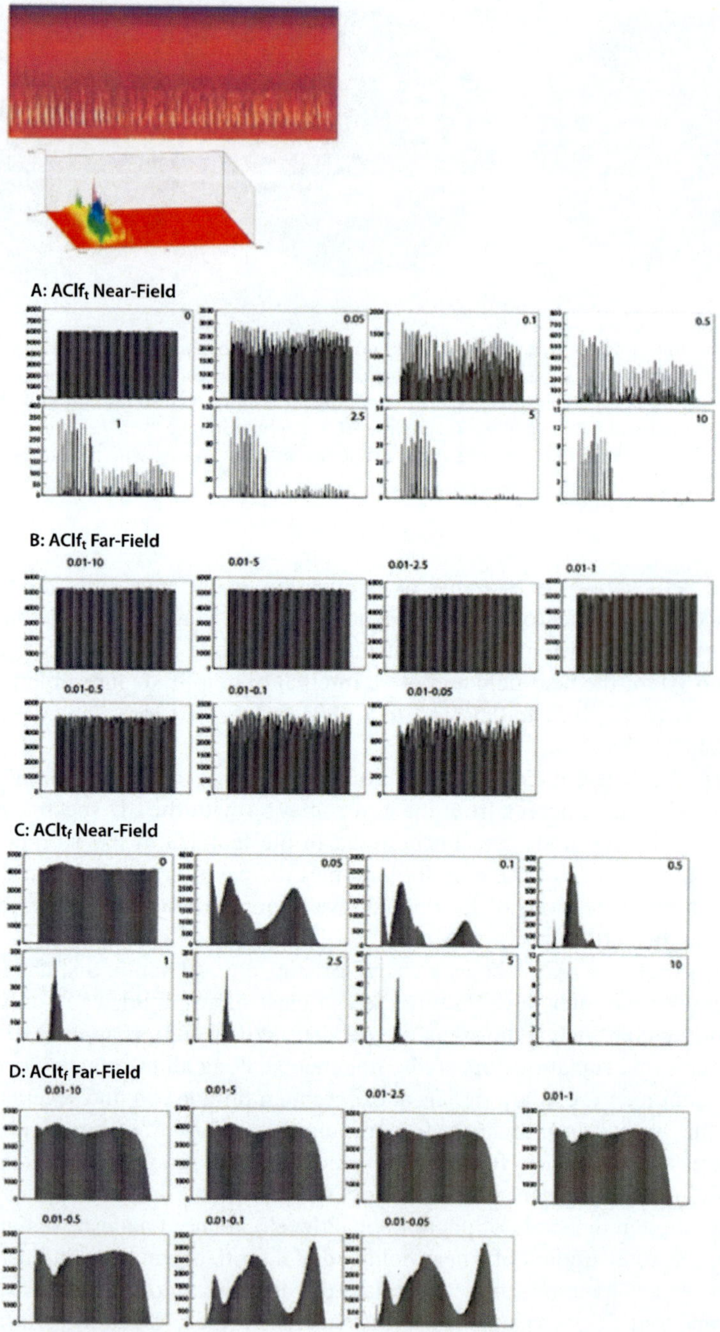

Fig. 2.19 A continuous sequence of chaffinch song with more individuals singing at different distance. Spectrogram and its 3D representation. (**A**) ACIf$_t$ obtained by fixing a threshold of near field from 0 to 10 of amplitude. (**B**) ACIf$_t$ obtained a threshold of far field in an interval of amplitude of 0.01 to 0.05, 0.01 to 0.1, 0.01 to 0.5, 0.01 to 1, 0.01 to 2.5, 0.01 to 5, and 0.01 to 10. (**C**) ACIt$_f$ obtained by fixing a threshold of near field from 0 to 10 of amplitude. (**D**) ACIt$_f$ obtained a threshold of far field in an interval of amplitude from 0.01 to 0.05, 0.1, 0.5, 1, 2.5, 5, and 10

2.9 Coding Soundscape Information

One of the main goals of ecoacoustics is to explore sounds and their patterns to better understand their role in the functioning of the ecological systems, across a broad range of functional, temporal, and spatial scales from population to landscapes. Moreover, it is not secondarily the necessity to find a relationship between soundscapes and structure, asset, and level of degradation of the ecological systems.

To better understand and interpret the acoustic environmental dynamics, the investigation of typology of acoustic patterns and their dynamics is central in ecoacoustic research. To achieve this goal, it is necessary to develop a robust methodology able to categorize the sonoscapes and to discover regularities/irregularities that govern the environmental processes.

The variation in sonoscape characters depends on many factors operating across a broad range of temporal and spatial scales, here a tentative list:

1. Ecoregion
2. Habitat
3. Position in the landscape
4. Climatic conditions
5. Weather conditions
6. Years
7. Seasonality
8. Daily hour

Although not easy to be perceived and measured, recurrent patterns and specific typologies characterize every sonoscape. When a sonoscape is represented in the frequency domain by a Fourier tranform, it appears like a sonic matrix in which sonic elements (signals) are more or less aggregated creating complex figures according to the typology of sound.

There are sonoscapes composed of very few signals, and others very rich, often intermediate conditions prevail (Fig. 2.20). Sonoscape appears in different status according to internal or external factors and causes where the occupation of the different spectral lines is due to the different typologies of sounds.

Localities poor of soniferous species have very simple and empty sonoscapes, whereas tropical primeval forests usually have sonoscapes very complex and full of signals because of the presence of thousands of soniferous species. Figure 2.21 shows a 3D representation of $ACIt_f$ from a Borneo and Amazon richest locations in terms of biophonies.

Inside the same localities, changes in sonoscape complexity occur according to the daily hours and the season. The recurrent patterns inside a temporal interval, for instance early morning, are affected by atmospheric and climatic conditions. In a windy or heavy rain days, biophonies are reduced for the masking/interference effect of wind and rain.

There are at least two typologies of problems when we search for categories of recurrent patterns inside a spectrogram: The first is the delimitation of each pattern,

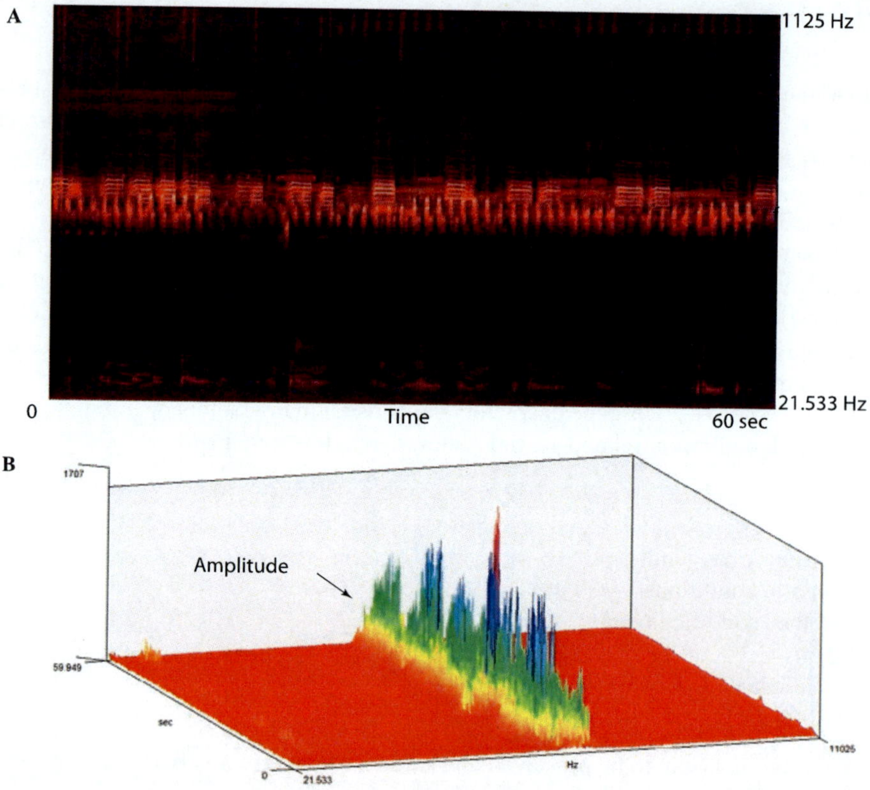

Fig. 2.20 A spectrogram (**A**) and its 3D (**B**) representation at intermediate signal density (at the medium frequencies) from a wav file of 60 s recorded in Madagascar on March 16, 2013, at 1:00 am (Courtesy of Anne Axel) characterized by a narrow spectral lines created by nocturnal insects

and the second is the minimum length of the temporal window to include distinct patterns.

The sonoscape produced by heavy rain can be distinguished easily from a bird chorus, and these differences can be quantified in a sonic matrix, but other conditions are not easy to be isolated. Here, we propose a mathematical approach utilizing a combination of ACI metrics, that transforms sonoscape patterns into a sequence of **acoustic codes** at which, *at posteriori*, we can empirically attribute specific ecoacoustic significance (ecoacoustic events) (Fig. 2.22) (Farina et al., 2016, 2018).

With a coding procedure, a sonic matrix is transformed into an ecosemiotic matrix where elements of the original matrix are aggregated and at which specific codes are assigned and interpreted according to an empirical knowledge.

The ecosemiotic matrix loses the spatial dimension transforming the spatial patterns observed into a sequence of codes. For this, we recognize three important conditions:

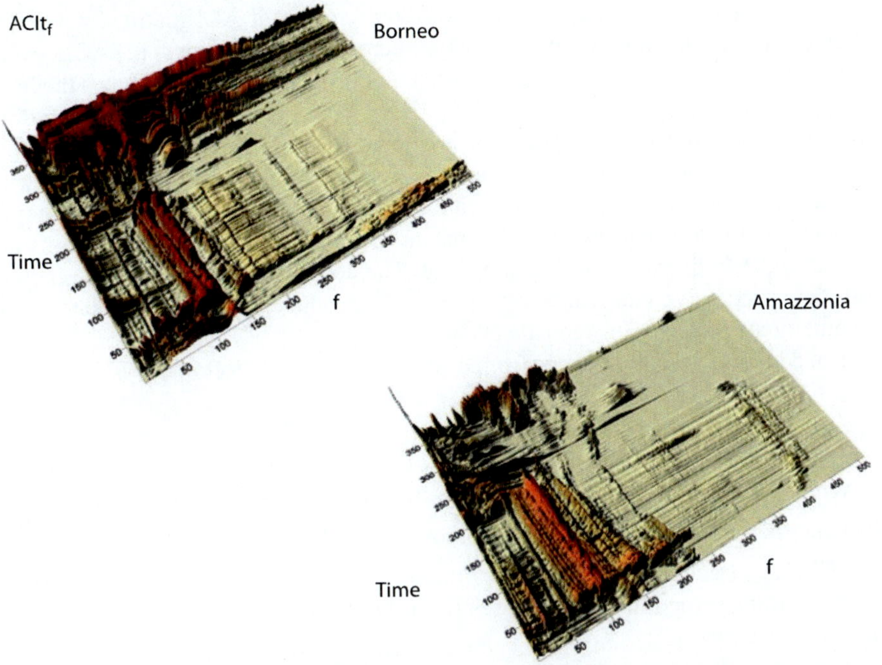

Fig. 2.21 The 3D distribution of 24 h of ACIt$_f$ from a Borneo and Amazon tropical forests. (Data from courtesy of David Monacchi)

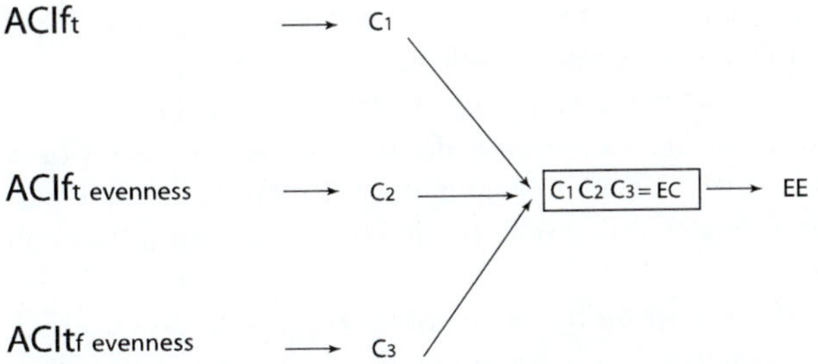

Fig. 2.22 The transformation of ACIf$_t$, ACIf$_{t\ evenness}$, and ACIt$_{f\ evenness}$ values in 10 categories of codes from 0 to 9 (C_1, C_2, C_3), and the combination of these three indices produces an ecoacoustic code EC = $C_1C_2C_3$ at which it is possible to utilize empirical evidence to attribute the meaning of an Ecoacoustic Event [EE]

1. The amount of information extracted.
2. The modalities with which the acoustic information is distributed along a temporal interval.
3. The modalities with which the acoustic information is distributed along each frequential line.

As consequence, we expect to find different values of ACI metrics, where [1] $ACIf_t$ calculates the amount of information across the frequencies, [2] $ACIf_t$ evenness shows the level of its equal distribution, and [3] $ACIt_f$ evenness shows the level of equipartition of the spectral lines $ACIt_f$. Every metric concurs to create three-digit codes where the most significant value is offered by $ACIf_t$ (Fig. 2.23).

For instance, heavy rain is expected to have the values of all the three metrics very high, because acoustic signal of rain is well represented with the same amplitude across all the frequencies and constant in time, and the spectral patterns are similar to the frequency plot of white noise. The combination of these three indices and their successive codification (from 0 to 9) produces a final code that can be eventually attributed to an acoustic event.

The equation to extract codes from a sonic matrix is the following:

$$EC = ACIf_t^{0,9} | ACIf_{t\,evenness}^{0,9} | ACIt_{f\,evenness}^{0,9}$$

Where $ACIf_t$ results from the normalization of $\sum_{t=1}^{n_t} (ACIf_t)$ for the entire set of data and then transformed into nine classes of abundance from 0 to 9.

$ACIf_{t\ evenness}$ and $ACIt_{f\ evenness}$ values that range from 0 to 1 (perfect equidistributed) are transformed into nine classes from 0 to 9, before being combined into an ecoacoustic code.

The sonic matrix is transformed in a unique three-digit code (f.i., 111, or 238, etc.) or in a number of codes as there are parts in which the sonic matrix has been scanned.

For instance in Fig. 2.24 is represented an example of conversion of the sonic matrix into codes according to a different resolution obtaining a number of codes corresponding to the number of the selected temporal steps.

Potentially, this procedure can return 1000 typologies of codes that are the result of the combination of these three code digits, ranging from 000 (this code is present only in theory and represents no sound at all) to 999 (saturated acoustic spectrum such as white noise, also this possibility looks remote, but heavy rains can produce a similar result).

The choice of the disposition of the three codes is based on some empirical evidences, but this sequence can be modified. For instance, it is possible to use $ACIt_f$ as first element of the code, and the two evenness can be inverted.

The ACI metric offers two possibilities: to modify internal resolution (grain size) and temporal scale of a sonic matrix.

The grain size is editable using the Clumping function that operates on the structure of the signals (see Sect. 2.5).

The choice of the temporal window at which to convert a sonic matrix into an ecosemiotic matrix remains challenging. In fact, after the change of the temporal resolution at which to subdivide the sonic matrix, we obtain a number of codes as the number of subdivisions.

Fig. 2.23 The combination of the three metrics ($ACIf_t$, $ACIf_{t\ evenness}$ and $ACIt_{f\ evennes}$ allows to classify every ecoacoustic event present in a sonic matrix. In this case using a ternary plot, we have represented the distribution of Ecoacoustic codes of (**A**) heavy rain and (**B**) dawn chorus of birds (Magliano, May 2, 2021, recording station #1, 44°14′ N, 10°03′ E)

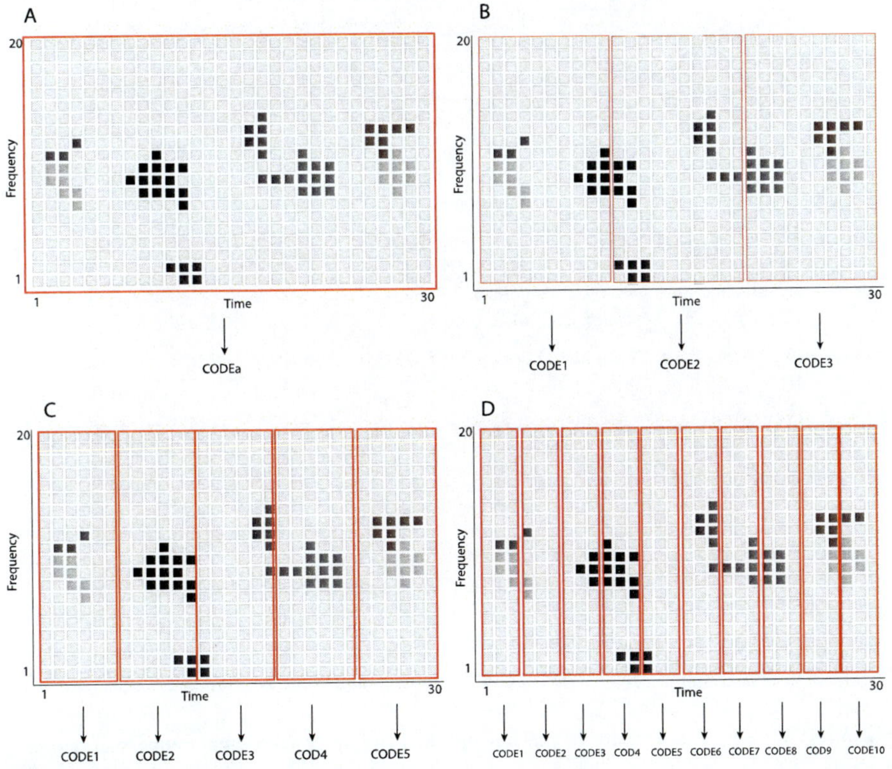

Fig. 2.24 Example of subdivision of a sonic matrix in submatrices in which a code computation is made. For each subdivision, we obtain a correspondent unique code

2.10 From Ecoacoustic Codes to Ecoacoustics Events (EE)

When a sonic matrix is transformed into an ecosemiotic matrix applying a coding procedure, it remains the difficult task to attribute a meaning to the ecoacoustic codes because only few are unambiguously representing a recognizable ecoacoustic event. For instance, heavy rain or a white noise has an ecoacoustic code of 999 (Fig. 2.25). However, the majority of codes can be used to describe a status more than an event.

To assure a mathematical treatment, the EE model is based on a sequence of temporal windows in order to capture potential signs according to the context and the interpreter.

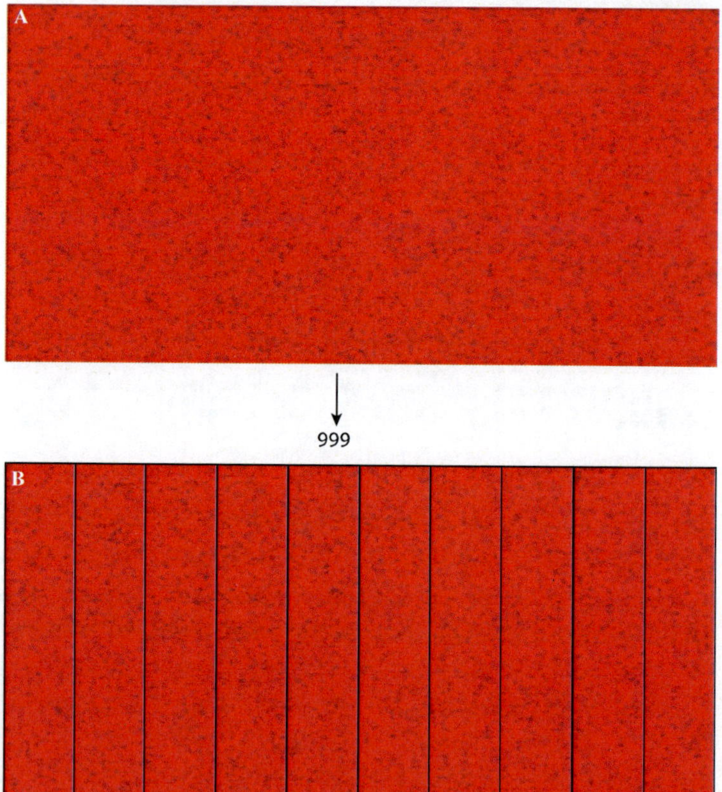

Fig. 2.25 White noise is a random signal having equal amplitude at different frequencies returning a constant value of ACI metrics at all scales. From figure **A** the ecoacoustic code that returns is 999, and this code is maintained at all the temporal scale selected (in the case of figure **B**, 10 temporal intervals of 4.5 s each)

The procedure to obtain an EE is a little complex because mathematically an EE is the combination of three codes obtained by the transformation of three ACI indices ($ACIf_t$, $ACIt_{f\ evenness}$, and $ACIf_{t\ evenness}$) into ten categories (from 0 to 9) and then their combination in the sequence $ACIf_t$, $ACIf_{t\ evennes}$, and $ACIt_{f\ evenness}$ produces a ternary code at which it's possible to attribute an ecoacoustic event.

The reasoning posed at the basis of the EE is represented by the potential discrimination of some emerging characters of a complex acoustic signal inside a sonic matrix.

Often, an EE is not coincident with the sound of a specific species but is a model with which it is possible to delimit a sonic coherence analyzing the entire set of the sonic information.

In addition, Ecoacoustic events can also be considered as emergent sonic patterns that are recognized by individual species during the completion of a specific function. In this case, the ecoacoustic events can be considered as acoustic eco-fields (Farina & Belgrano, 2004, 2006).

To clarify the meaning of EE codes, we present three examples:

1. A code of 199 indicates a low amplitude sequence (**1**99), with even temporal distribution (1**9**9), and a broad distribution of information across frequencies (19**9**).
2. A code of 191 indicates a low amplitude sequence (**1**91), with even temporal distribution (1**9**1), and a narrow frequency distribution (19**1**).
3. A code of 119 indicates a low amplitude sequence (**1**19), with narrow temporal distribution (1**1**9), and a broad distribution of information across frequencies (11**9**).

2.11 Scaling a Sonic Matrix to Obtain Ecoacoustic Codes EC

The temporal scale at which to operate and the level of aggregation are the result of an empiric approach. This approximation creates difficulties in the standardization of the ecoacoustics methodologies. A strategy to escape this uncertainty is to operate to a multiscale, which means to change the temporal resolution and successively to observe the behavior of the values obtained at different temporal resolution.

In order to explore the internal complexity of a sonic matrix, we can either operate on the dimension of the temporal interval in which to subdivide the matrix or act on the clumping. In the first case, we change the temporal extension on which to extract Ecoacoustic Codes (EC), and in the second case, we operate on the grain resolution of the matrix. In the last case, the data are aggregated before the application of the ACI metrics as reported in Fig. 2.26. See in Fig. 2.27 to 2.29 examples of temporal scale of Ecoacoustic Codes.

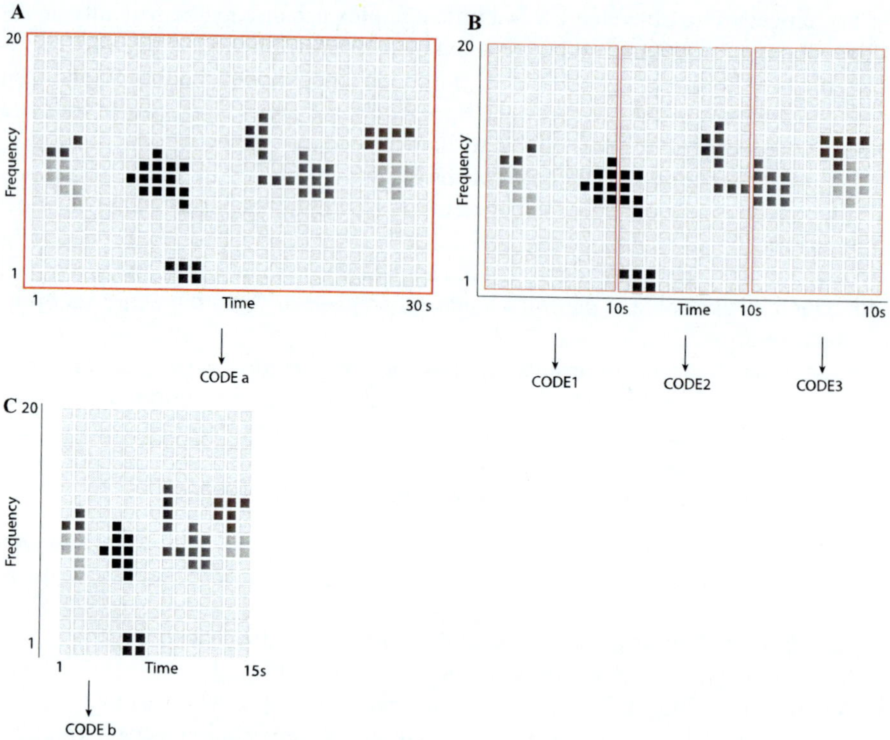

Fig. 2.26 In this example, a unique code is extracted by a sonic matrix when we operate at a resolution of 30 s. Operating at different temporal scale means to subdivide the matrix in submatrices of equal length (**A**) and a code is obtained for each temporal interval (**B**), in this case, the interval was of 10 s. But, if we apply a clumping value =2, we obtain a matrix that is half of the size of the original (**C**) and a unique code from this new matrix

With the first approach, a sonic matrix can be subdivided at different temporal windows. For instance, if the analysis of a sonic matrix is done on the entire matrix, this approach returns only one ecoacoustic code. But if for example we subdivide the matrix in 10 parts, this returns 10 codes that are spatially coherent with original file.

When we apply the aggregation, that is the clumping procedure already described in Sect. 2.5, the level of aggregation returns a matrix that is the half of original if the clumping is 2, or forth if clumping is 4 and so on (Fig. 2.30).

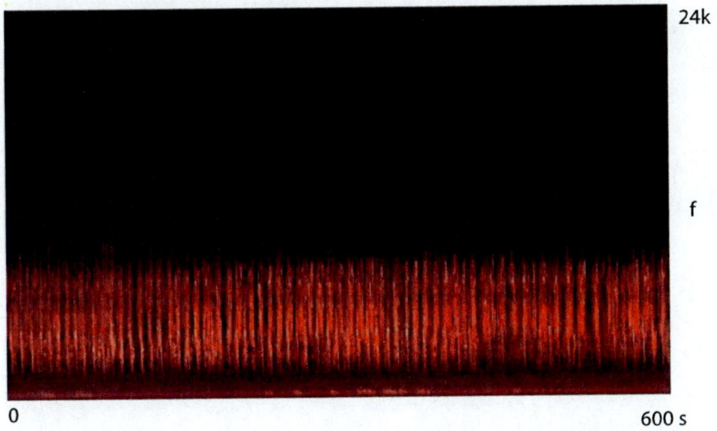

Scale	600	300	200	100	50
	941	931	731	531	420
		942	941	831	631
			632	941	631
				541	941
				632	851
					951
					941
					642
					531
					432
					741
Tot SC	1	2	3	5	10

Fig. 2.27 An example of dawn chorus. Codes have been extracted at the temporal resolution of 600, 300, 200, 100, and 50 second. According to this resolution, the number of maximum expected codes (richness) should be 1, 2, 3, 6, and 12. In effect, we obtained 1, 2, 3, 5 and 10 uniqe codes

2.12 Ecoacoustic Code Diversity $\left(H'_{EC}\right)$

Once time that we have obtained the Ecoacoustic Codes, this metric can be used to calculate the **Diversity of Ecoacoustic Code** $\left(\mathbf{H}'_{\mathbf{ec}}\right)$ (Shannon & Weaver, 1949; Farina et al., 2018)

$$H'_{EC} = -\sum e_i \log e_i$$

Where *ei* is the relative abundance of each *EC* in the entire collection (all data aggregated for each sampling location or time).

This index, which is extremely synthetic and informative, proves important when we compare one location with another, or different sonotopes from the same locality but at different hour of the day. The variations in sonic patterns are well represented

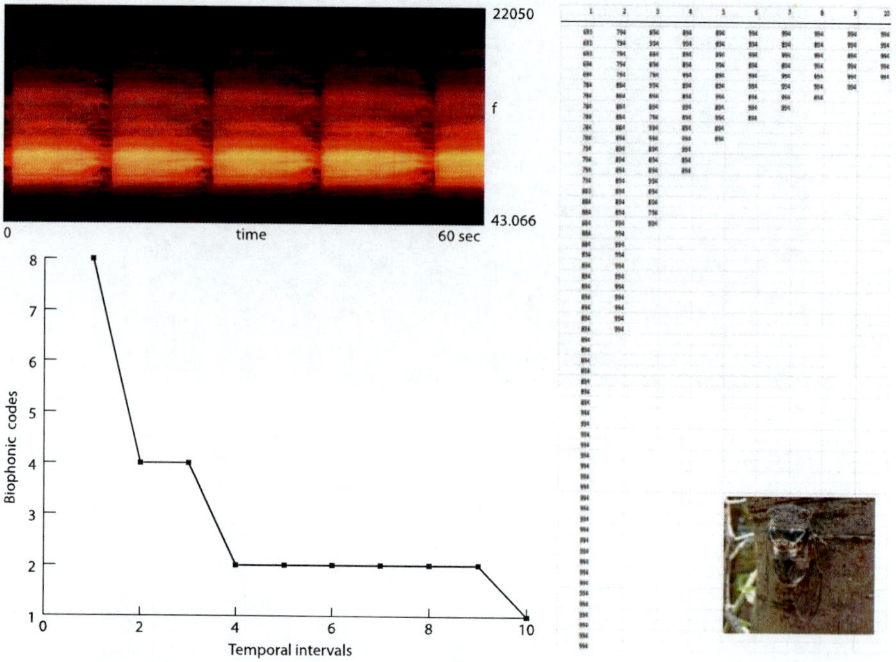

Fig. 2.28 Relationship between number of codes and temporal scaling of the song of a cicada (*Lyristes plebejus*) (Courtesy of Jerome Sueur, NMHN, Paris). A file of 60 second has been analyzed at 10 temporal scale from 1 to 10 second, resulting 88 codes at 1 second, 4 at 2 and 3 second, 2 codes from 4 to 9 second, and finally one code at 1 second

in the EC because this code represents a combination of three relevant indicators of the sonic complexity. The calculus of the diversity emphasizes the sonic complexity inside a single sound file.

Of course the application of this index, which in the **SonoScape** software is called *ee_entropy*, requires that some conditions are well established and that the research project is well organized.

In particular, this entropy can be applied to the collection of codes that are found at each temporal interval in which the sound file is subdivided. For instance, if we have a file of 300 second and we subdivide the files in intervals of 10 second each, we will have 30 codes. At this collection of 30 codes, we will apply the H'_{EC} and for each file, we obtain a unique value.

In this way, the internal complexity of a file is represented by a single numerical expression that takes into account the number of codes and their relative importance. This approach allows a comparison between different sonic files because diversity does not depend on the absolute values of the components (the temporal subdivisions) but on the relative variation in values among the entire collection of matrix segments. In Fig. 2.31 is reported an example of a day recording at Agnolo

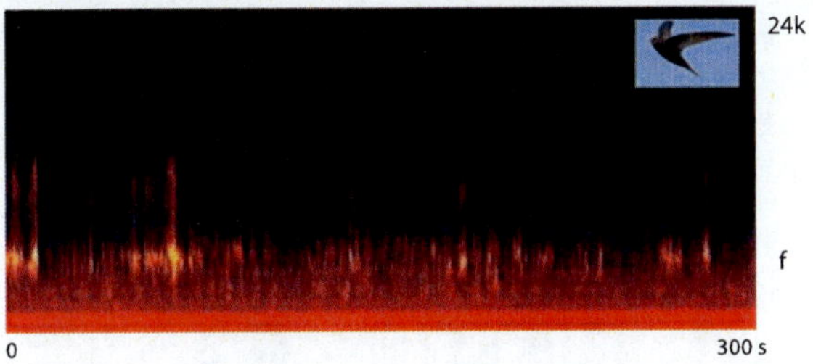

10	20	30	40	50	60	90
730	981	971	981	981	981	981
520	681	981	981	981	881	781
30	881	981	690	781	881	781
120	981	690	781	881	781	
10	691	881	781	681	781	
330	590	791	681	781		
930	781	881	681			
130	691	681				
30	681	781				
30	781	781				
30	681					
30	580					
130	590					
220	781					
130	681					
140						
30						
120						
220						
120						
120						
110						
20						
120						
20						
30						
130						
110						
110						
20						
12	7	6	4	4	3	2

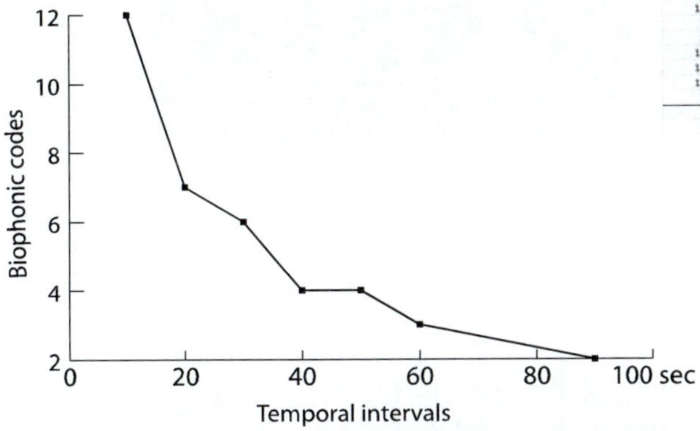

Fig. 2.29 Relationship between number of codes and temporal scaling. Soundscape in an urban context where swifts (*Apus apus*) are dominating in some hours of day (Fivizzano, June 21, 2020, 08:18 a.m., 44°14′ N, 10°07′ E)

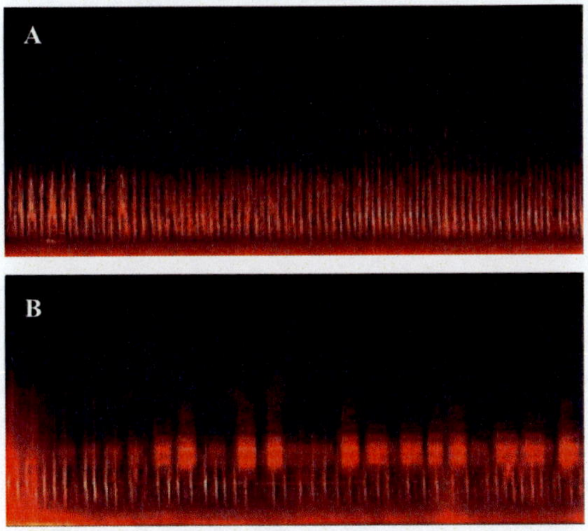

Clumping	Ecoacoustic code	
	A	**B**
0	**931**	**922**
2	**921**	**922**
3	**962**	**962**
4	**932**	**932**
5	**972**	**973**
6	**942**	**943**
7	**942**	**943**
8	**982**	**943**
9	**942**	**993**
10	**943**	**943**
20	**943**	**943**
30	**943**	**943**
40	**943**	**944**
100	**944**	**844**

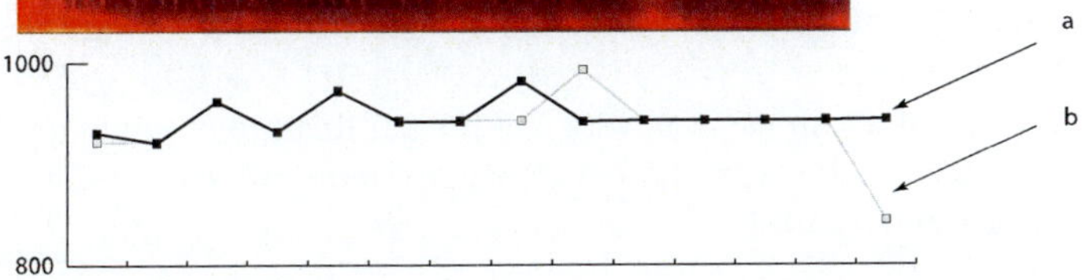

Fig. 2.30 Variation in clumping value during the assessment of ecoacoustic code in two periods in Sasso Fratino (43° 11′N, 11°47′ E) (Courtesy of Gianni Pavan) (**A**) 06:00 a.m.; (**B**) 01:00 p.m. of 10 min each. At every clumping, the files are temporally reduced of 1/2, 1/3... 1/100

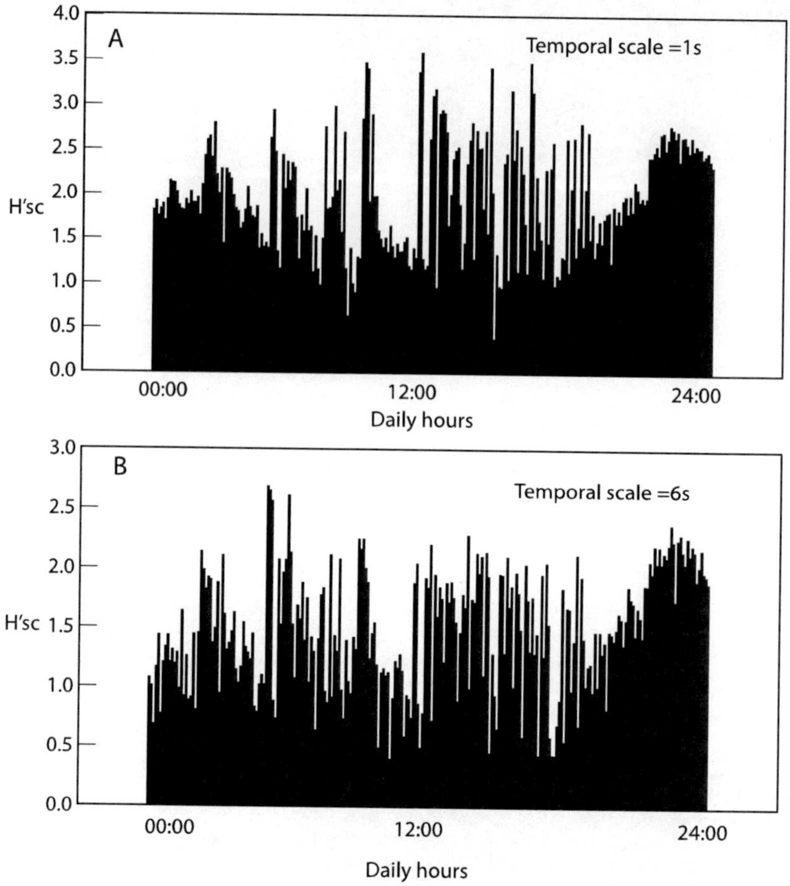

Fig. 2.31 H'_{EC} along an entire day at Agnolo lake (44°13′, 10°04′ E) recording station calculated at the temporal resolution of 1 and 6 s, respectively

lake collecting 240 files each 5 minutes in length. Every minute has been subdivided into temporal segments from 1 to 6 s. The H'_{EC} of each 5 minutes has been plotted along the entire day.

This metric can also be used to evaluate the effect of the temporal scaling. In fact, it is possible to calculate H'_{EC} at each temporal scale selected and to extract the fractal dimension (section below).

The internal complexity of a rain event is very low, and if this file is subdivided into subfiles, every subdivision returns the same value; consequently, H'_{EC} should be close to 0 and its evenness equal to 1.

2.13 Sonic Signature Dissimilarity (SSD)

The dissimilarity index (Orloci, 1967; Legendre & Gallagher, 2001) offers the possibility to investigate the level of distance between successive sonic signatures (ACIt_f) (Farina et al., 2018).

$$\text{SSD} = \frac{2}{n^2} \sum_{i=j}^{n} \sum_{j=1}^{n} \text{Dchord}\left(x_i, x_j\right)$$

where x_i and x_j are the successive processing steps; and n is the total number of SS which varies depending on the temporal aggregation parameter (i.e., for temporal aggregation 1 s, $n = 600$); the function $\text{Dchord}(x_i, x_j)$ is formulated by:

$$\text{Dchord}\left(x_i, x_j\right) = \sqrt{\sum_{f=1}^{n_f} \left(\frac{y_{if}}{\sqrt{\sum_{f=1}^{n_f} y_{if}^2}} - \frac{y_{jf}}{\sqrt{\sum_{f=1}^{n_f} y_{jf}^2}} \right)^2}$$

where y_{if} and y_{jf} are the elements of the i_{th} and j_{th} ACIt_f, and n_f is the number of spectral lines (in this case 512). The chord distance is a maximum of $\sqrt{2}$ when the sonic signature of two sampling units is completely different and is 0 in the case of an identical distribution (Fig. 2.32).

2.14 The Fractal Dimension of the Ecoacoustic Codes (D_{EC}) and of Sonic Signature Dissimilarity (D_{SSD})

Fractal mathematics was applied to the ECs, calculated using the box-counting method for fractal dimension (Mandelbrot, 1983; Feder, 1988; Li et al., 2009), where the logarithm of the number of ECs was regressed with the logarithm of the scale so as to calculate the fractal dimension (Monacchi & Farina, 2019). Fractal dimension D is defined as a ratio that provides a statistical index of complexity. It compares how the detail in a given pattern changes with the scale at which it is measured and where

$$D = 1\text{-}b$$

where b is the slope of the regression equation between the logarithm of the scale of temporal processing (f.i., 1, 5, 10, 15, 20 s) and the logarithm of number of ECs and of the values of SSD, respectively. D ranges from 1 to 2. D is equal to 1 when the number of ECs or *SSD* does not change across the scale. Higher D values are rendered if details (number of ECs) emerge quickly as the observational scale decreases, demonstrating thus a more complex pattern. D is not a proxy for biodiversity, but is

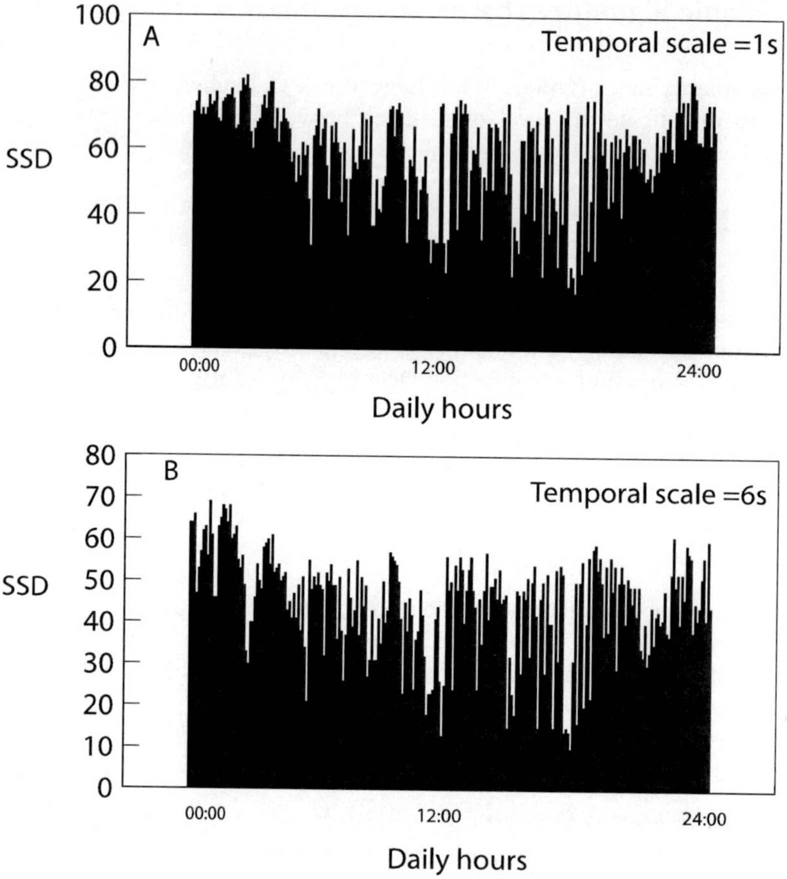

Fig. 2.32 Example of sonic signature dissimilarity calculated for the day at Agnolo lake (44°13′, 10°04′ E) recording station and based on temporal steps of 5 min subdivided into 1 and 6 s respectively

an indicator of sonic complexity. A high value of *D* may be obtained if several species are present with a narrow spectral frequency (the majority of insects). However, where acoustic communities have species that use a broad spectral frequency, such as songbirds in temperate biomes, *D* may be high despite the fact that such communities are not rich in species.

References

Bertucci, F., Guerra, A. S., Sturny, V., Blin, E., Sang, G. T., & Lecchini, D. (2020). A preliminary acoustic evaluation of three sites in the lagoon of Bora Bora, French Polynesia. *Environmental Biology of Fishes, 103*(8), 891–902.

Bertucci, F., Parmentier, E., Lecellier, G., Hawkins, A. D., & Lecchini, D. (2016). Acoustic indices provide information on the status of coral reefs: An example from Moorea Island in the South Pacific. *Scientific Reports, 6*, 33326.

Bittencourt, L., Barbosa, M., Bisi, T. L., Lailson-Brito, J., Jr., & Azevedo, A. F. (2020). Anthropogenic noise influences on marine soundscape variability across coastal areas. *Marine Pollution Bulletin, 160*, 111648.

Bolgan, M., Amorim, M. C. P., Fonseca, P. J., Di Iorio, L., & Parmentier, E. (2018). Acoustic complexity of vocal fish communities: A field and controlled validation. *Scientific Reports, 8*(1), 1–11.

Brumm, H. (2009). Song amplitude and body size in birds. *Behavioral Ecology and Sociobiology, 63*(8), 1157–1165.

Ceraulo, M., Papale, E., Caruso, F., Filiciotto, F., Grammauta, R., Parisi, I., Mazzola, S., Farin, A., & Buscaino, G. (2018). Acoustic comparison of a patchy Mediterranean shallow water seascape: Posidonia oceanica Meadow and sandy bottom habitats. *Ecological Indicators, 85*, 1030–1043.

Colonna, J. G., Carvalho, J. R., & Rosso, O. A. (2020). Estimating ecoacoustic activity in the Amazon rainforest through information theory quantifiers. *PLoS One, 15*(7), e0229425.

Davies, B. F., Attrill, M. J., Holmes, L., Rees, A., Witt, M. J., & Sheehan, E. V. (2020). Acoustic complexity index to assess benthic biodiversity of a partially protected area in the southwest of the UK. *Ecological Indicators, 111*, 106019.

Duarte, M. H. L., Sousa-Lima, R. S., Young, R. J., Farina, A., Vasconcelos, M., Rodrigues, M., & Pieretti, N. (2015). The impact of noise from open-cast mining on Atlantic forest biophony. *Biological Conservation, 191*, 623–631.

Farina, A. (2014). *Soundscape ecology: Principles, patterns, methods and applications*. Springer.

Farina, A., & Belgrano, A. (2004). The eco-field: A new paradigm for landscape ecology. *Ecological Research, 19*(1), 107–110.

Farina, A., & Belgrano, A. (2006). The eco-field hypothesis: Toward a cognitive landscape. *Landscape Ecology, 21*(1), 5–17.

Farina, A., Gage, S. H., & Salutari, P. (2018). Testing the ecoacoustics event detection and identification (EEDI) approach on Mediterranean soundscapes. *Ecological Indicators, 85*, 698–715.

Farina, A., & Pieretti, N. (2014). Sonic environment and vegetation structure: A methodological approach for a soundscape analysis of a Mediterranean maqui. *Ecological Informatics, 21*, 120–132.

Farina, A., Pieretti, N., Salutari, P., Tognari, E., & Lombardi, A. (2016). The application of the acoustic complexity indices (ACI) to ecoacoustic event detection and identification (EEDI) modeling. *Biosemiotics, 9*(2), 227–246.

Feder, J. (1988). *Fractals*. Plenum Press.

Fuller, S., Axel, A. C., Tucker, D., & Gage, S. H. (2015). Connecting soundscape to landscape: Which acoustic index best describes landscape configuration? *Ecological Indicators, 58*, 207–215.

Gasc, A., Sueur, J., Pavoine, S., Pellens, R., & Grandcolas, P. (2013). Biodiversity sampling using a global acoustic approach: Contrasting sites with microendemics in New Caledonia. *PLoS One, 8*(5), e65311.

Hurlbert, S. H. (1978). The measurement of niche overlap and some relatives. *Ecology, 59*, 66–77.

Legendre, P., & Gallagher, E. D. (2001). Ecologically meaningful transformations for ordination of species data. *Oecologia, 129*, 271–280.

Levins, R. (1968). *Evolution in changing environments. Some theoretical explorations*. Princeton University Press.

Li, J., Du, Q., & Sun, C. (2009). An improved box-counting method for image fractal dimension estimation. *Pattern Recognition, 42*(11), 2460–2469.

Mandelbrot, B. B. (1983). *The fractal geometry of nature*. Freeman.

Mingers, J., & Standing, C. (2018). What is information? Toward a theory of information as objective and veridical. *Journal of Information Technology, 33*(2), 85–104.

Monacchi, D., & Farina, A. (2019). A multiscale approach to investigate the biosemiotic complexity of two acoustic communities in primary forests with high ecosystem integrity recorded with 3D sound technologies. *Biosemiotics, 12*(2), 329–347.

Morri, D. (2006). *L'eco-field, ipotesi ed applicazioni*. PhD dissertation, Urbino University.

Orloci, L. (1967). An agglomerative method for classification of plant communities. *Journal of Ecology, 29*(1), 193–206.

Pieretti, N., Farina, A., & Morri, D. (2011). A new methodology to infer the singing activity of an avian community: The Acoustic Complexity Index (ACI). *Ecological Indicators, 11*(3), 868–873.

Scott, J. M., Ramsey, F. L., & Kepler, C. B. (1981). Distance estimation as a variable in estimating bird numbers from vocalizations. *Studies in Avian Biology, 6*, 334–340.

Shannon, C. E., & Weaver, W. (1949). *The mathematical theory of communication*. University of Illinois Press.

Stonier, T. (1996). Information as a basic property of the universe. *Biosystems, 38*, 135–140.

Chapter 3
Introduction to the SonoScape, an Open-Source Software Application in MATLAB®

Abstract SonoScape is a new software to process wav files using a full set of Acoustic Complexity metrics [the ACI metrics ($ACIf_t$, $ACIt_f$) and derivative indices: $ACIf_t$ *evenness*, $ACIt_f$ *evenness*, *Sonic Signature Dissimilarity* (SSD), *Ecoacoustic Events* (EE), *EE Entropy*, and *Fractal dimensions* of EE and SSD], developed under the MATLAB® platform. SonoScape is enriched of several options to manage single and large collection of sonic files. It enables a hierarchical structure of file management and fully automates the ecoacoustic analyses of sonic files stored across different folders/layers on the disk, optimally and maximally reducing human efforts in the repetitive manual operations on each single file or folder. This possibility allows automatic computation of files belonging to different days and different recording stations, or different seasons inside the same recording station. SonoScape can operate at different clumping and temporal scales. An energy filter can be applied along a range of values returning information on the importance of signals inside a near or a far field. Exclusion of some frequencies in the calculation can also be obtained through the frequency filter parameter. The SonoScape yields human-readable results for each individual acoustic signal as well as summary files to ease subsequent analyses including statistical analyses. It returns analytic results and their aggregations according to different configurations. It also offers functions to visualize the results using customized 3-D plots or ternary plots, intuitively demonstrating the patterns of ACIs based on the vast number of numerical results. SonoScape provides utilities including the export of the FFT matrix for each wav file, the export of EE ternary plots for each folder, the sorting of wav files according to hours and days, and the setting of Max value of $ACIf_t$ for encoding EE.

Keywords SonoScape · Installation instructions · Software configuration · Acoustic Complexity Index processing

A. Farina, P. Li, *Methods in Ecoacoustics*, Frontiers in Ecoacoustics 1,
https://doi.org/10.1007/978-3-030-82177-7_3

3.1 About the SonoScape Open-Source Software Project

Despite recent advances in ecoacoustic analysis (f.i., Farina & Salutari, 2016; Villanueva-Rivera & Pijanowski, 2018; Sueur et al., 2020), difficulties persist to manage data friendly and speedily. On the one hand, this encourages researchers in the acoustic field to collaborate with experts in other disciplines such as mathematics, signal processing, statistical physics, and nonlinear dynamics. On the other hand, the situation has deeply motivated the development of software applications that are dedicated to the ecoacoustic field.

We thus initiated the SonoScape open-source software project in the spring of 2020 which has so far been going smoothly. The SonoScape is developed under the MATLAB® platform. At the date of writing up this chapter, we have released the first version (ver. 1.1.0426) of this software application. The SonoScape software is openly available at Zenodo and Github <https://github.com/pliphd/SonoScape> (Li & Farina, 2021).

3.2 Competencies

SonoScape processes audio data written in Waveform Audio File format [wav]. It computes the ACI metrics ($ACIf_t$, $ACIt_f$) and derivative indices like $ACIf_{t\ evenness}$, $ACIt_{f\ evenness}$, Sonic Signature Dissimilarity (SSD), Ecoacoustic Events Richness (EE), EE Entropy, and Fractal dimensions of EE and SSD (Fig. 3.1). Graphical representations of data (ACI visualizer and Ecoacoustic Event visualizer) are also provided. Utilities can help:

(a) Save intermediate data for further processes, for example the FFT matrix for each file.
(b) Manage the data, for example, sorting data from different recording methodologies and purposes, creating specific folders in which to transfer files homogeneous for some characters, for instance for hour or for day.
(c) Manage to apply fixed Max $ACIf_t$ throughout the analyses to standardize Ecoacoustic events or to adaptively calculate its values.

3.3 Installation of the SonoScape Software Application

3.3.1 System Requirements

The SonoScape will run on 64-bit Macintosh or Windows 8/10 computers. At least 2 Gigabytes hard disk space is recommended to install the application, and 8 Gigabytes RAM is recommended in case of working with large ecoacoustic data sets.

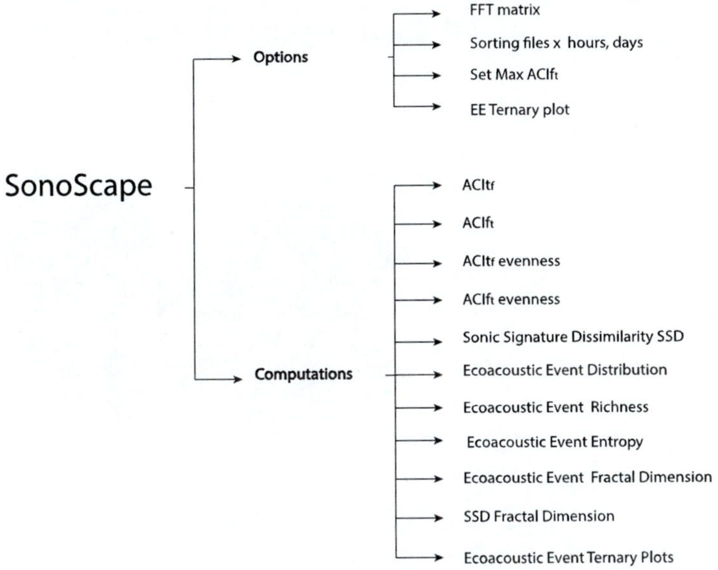

Fig. 3.1 Options and computations provided by SonoScape

The standalone installation of the SonoScape does not depend on the existing installation of MATLAB®, but the MATLAB® command line installation does. In the latter case, MATLAB® version R2020b (the latest version when we were writing this book) or newer releases are recommended in order for the SonoScape to function properly. In addition, the MATLAB® Signal Processing Toolbox (version 8.5 or later releases) and the MATLAB® Statistical and Machine Learning Toolbox (version 12.0 or later releases) are required for the MATLAB® command line installation of the SonoScape.

3.3.2 Standalone Installation of the SonoScape

Here, we demonstrate the standalone installation of the SonoScape in a 64-bit Windows 10 laptop. Network connection is needed since the standalone installation package will request the MATLAB® Runtime from remote servers (of the MathWorks, Inc.). Depending on the speed of the network connection, the installation may take a while since the MATLAB® Runtime is about 1 Gigabyte in size.

The installation wizard will run with a double-click on the installer "SonoScapeInstaller_web.exe". Figure 3.2 shows the boot interface of the wizard.

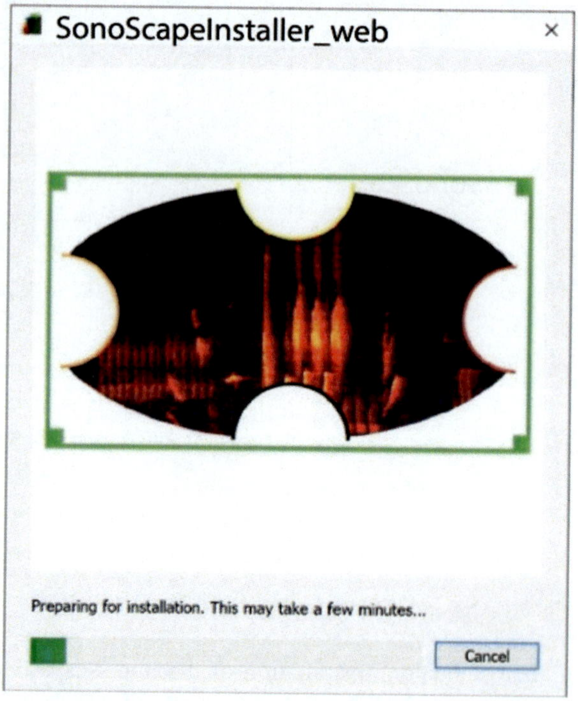

Fig. 3.2 SonoScape installation wizard. Shown the boot interface

This interface may disappear in seconds or minutes depending on the configuration of the computer.

After this step, the installation wizard will guide the user to choose the location to install the application. By default, this location should be a folder named "SonoScape" under the "Program Files" folder in the root directory (see Fig. 3.3).

An extra step after clicking "Next >" button is the installation of dependent software, the MATLAB® Runtime in this case. The default location for installing it is "MATLAB\MATLAB Runtime" in the "Program Files" folder in root directory (Fig. 3.4).

After setting the location for MATLAB® Runtime, the installation wizard will pop up the license agreement for using MATLAB® Runtime (Fig. 3.5), which is a standard agreement to grant personal, noncommercial use. Users need to accept this agreement in order to move forward with next step, which shows the confirmation information including where the SonoScape will be installed, where the MATLAB® Runtime will be installed, and the download size (Fig. 3.6). But clicking "Install", the installation wizard will show the progression of downloading and installation with a progress bar.

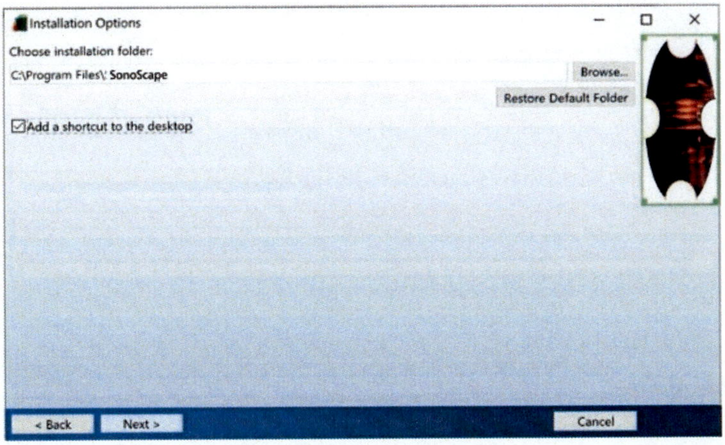

Fig. 3.3 SonoScape installation wizard. The installation options interface is shown in this figure. Check the checkbox to add a shortcut to desktop. Installation folder can be changed by clicking the "Brower…" button or restored by clicking the "Restore Default Folder" button. Click "Next >" for next step

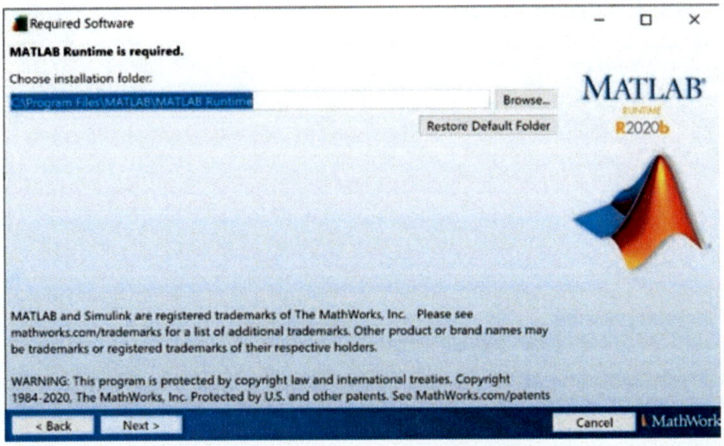

Fig. 3.4 SonoScape installation wizard. The MATLAB® Runtime interface is shown in this figure

3.3.3 MATLAB® Command Line Installation of the SonoScape

Two files are required to execute this installation: *Install_SonoScape.m* and *SonoScape_1.1.0426.zip*.

Start MATLAB®. Set the working directory to the folder where the "*Install_SonoScape.m*" and "*SonoScape_0.1.0426.zip*" files are located. In MATLAB®, the working directory means the current location on disk that MATLAB is currently working at. It shows in the address bar right on top of the Command Window. It can

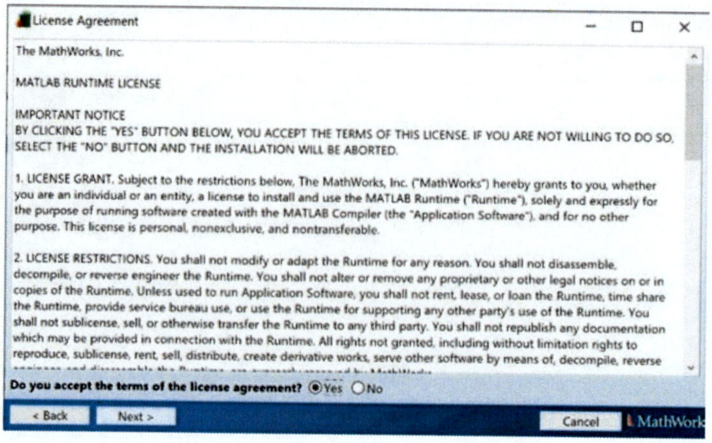

Fig. 3.5 SonoScape installation wizard. The license agreement interface is shown in the figure

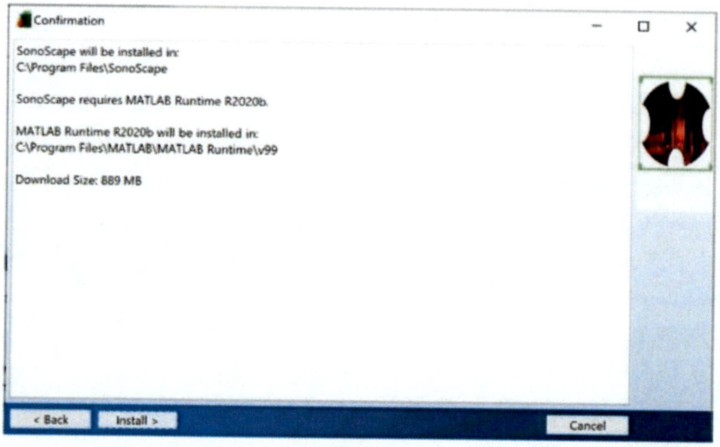

Fig. 3.6 SonoScape installation wizard. The confirmation page is shown in figure

be changed using the set of browser buttons or using command "cd the location that users want to set as the current working folder", for example, cd C:\Downloads\SonoScape.

Type "Install_SonoScape" on MATLAB® command window. The installation process will direct you to select which folder you would like to install the application. By default, it will be the current working directory. Successful prompt will be shown on MATLAB® command window if the installation goes through. The installation will also remove the SonoScape installed before (if there is one). This usually happens when the users update the SonoScape software.

3.3.4 *Registration*

Registration is not required after installing the application, since the SonoScape is an open-source software application under the GNU General Public License v3.0.

3.4 The User Interface of the SonoScape

To run the SonoScape, there are three possibilities:

(a) Double-click the icon on desktop (for standalone installation).
(b) Choose the SonoScape App from the Start menu of Windows 8/10 (for standalone installation).
(c) Type "SonoScape" on MATLAB® Command Window (for MATLAB® command line installation).

Figure 3.7 shows a snapshot of the main user interface of this software application. There are mainly five sections on this window:

1. Menu bar
2. Folder list
3. File list
4. Status bar
5. ACI computation
6. Post-ACI computation

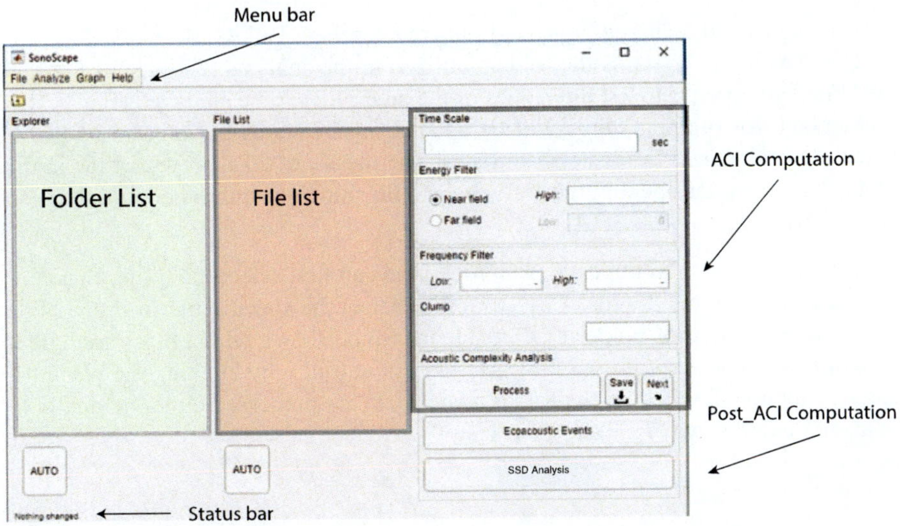

Fig. 3.7 The main interface of the SonoScape application and the submenu items

File

Open
Sort
Options
Exit

Analyzing

Pattern recognition

Graph

ACI Visualizer
Ecoacoustic Event Visualizer (Ternary Plot)

Help

About

Fig. 3.8 The menus and their submenus: "File" is composed of "Open Folder" (select wav files); "Sort" individual files according daily folders; "Options": "Import" (import gauge time), "Save" (select what files to save and define ASCII file separator), "Compute" (fixed or adaptive ACIf$_t$ max); and "Exit" commands. "Analyzing" performs "Pattern recognition". "Graph" allows the "ACI Visualizer" and the "Ecoacoustics Event Visualizer". "Help" contains the basic information about the release and authors

In this first release of the SonoScape, there are four Menu bar items:

(a) File
(b) Analyze
(c) Graph
(d) Help

The menus and their submenus are illustrated in Fig. 3.8.

The "Explorer" contains the folders utilized in the analysis ("Folder List"), and the "File List" area is to list the names of the audio files to be analyzed (Fig. 3.9).

Message for running status will be updated in the status bar area in real time (examples are shown in Figs. 3.7 and 3.9). The message "*Ready*" will show in the Status bar area after opening the software and finishing the automated initialization on background.

"Importing Files" To import a .wav file, users can first select the folder that contains the file from the "Folder List", and then select the specific file from the "File List". Note that in the "Folder List", the icon will be shown with a black triangle if there are subfolders within the folder. Users can unfold it by clicking the black triangle or by double-clicking the folder name. Data are processed either manually or in an automatic way (Fig. 3.9).

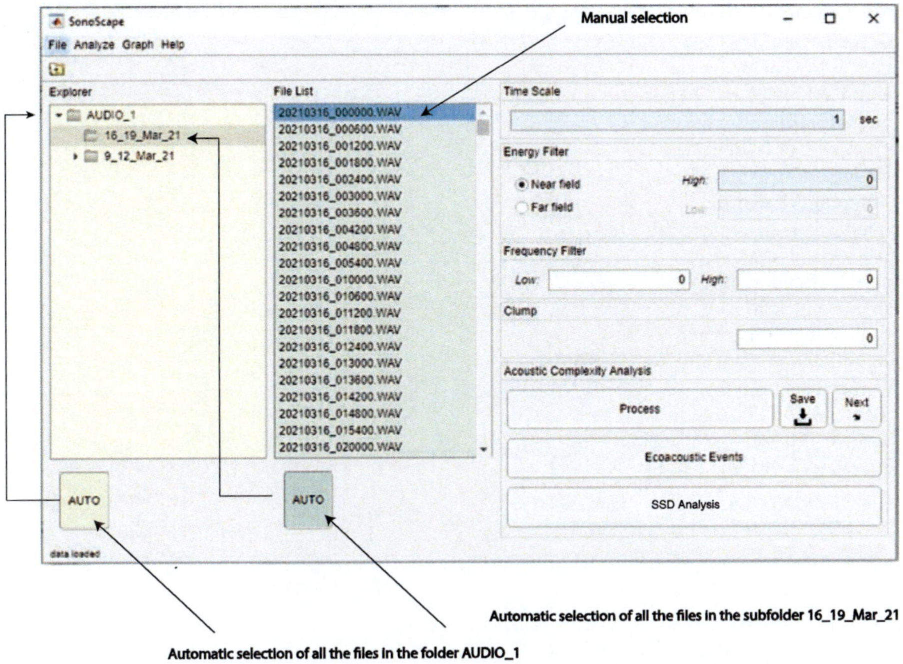

Fig. 3.9 Main panel in which in the "Explorer" are listed the folders and in the "File List" are indicated the files that will be processed separately after manual selection or automatically

3.4.1 Input of Parameters

Parameters for calculating ACI can be defined in the ACI area (Figs. 3.7 and 3.10).

These parameters include:

"Time Scale" Every file is subdivided according to the time scale. For instance, for a file of 60 seconds, a time scale of 1 means to obtain 60 elaborations of 1 s each. If set to 5 s, we obtain 12 elaborations, and so on.

There are two options to choose from—"Near field" and "Far field".

"High" [Near field]: this parameter, empirically fixed, excludes from the ACI calculation every element of the sonic matrix that is > the fixed value.

This subroutine is useful to exclude from computation high intensity sounds like a thunder or the passage of an airplane, or very loud sound that could degrade the EE calculus.

"Low" [Far field]: This parameter, empirically fixed, excludes from the ACI calculation every element of the sonic matrix < the fixed value. This subroutine excludes faint sounds.

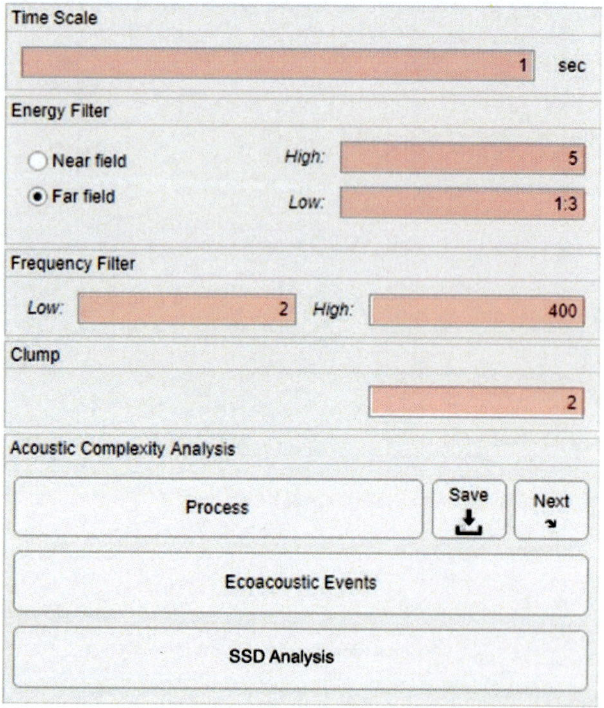

Fig. 3.10 Example of parameter setting: Time scale = 1; Near field: High = 5; Low = 1:3; Frequency filter: processing of frequencies >2 and < 400 spectral line; Clumping =2. For the energy filter, since there is a vectorized input (1:3), the calculation will be done using each combination of Low and High filters. Specifically, the elements used for calculation will be within [1, 5], [2, 5], and [3, 5], separately

"Low frequency filter" (LFF): This parameter excludes from ACI computation a number of frequencies equal to the input value. For instance, if LFF = 4, it would exclude the first four spectral lines.

"High frequency filter" (HFF): This parameter excludes from ACI computation a number of frequencies higher than a threshold. For instance, if HFF = 10, it means that all the frequencies higher than the spectral line 10 are excluded.

"Time scale", "High" and "Low energy filters", "Near field", "Far field", "Low frequency", and "High frequency filter" all accept vectorized inputs. For example, if users want to perform the analysis under time scales from 1 second all through to 10 second, they can use either [1, 2, 3, 4, 5, 6, 7, 8, 9, 10] in this edit box, or simply 1:10, which means to generate a vector from 1 to 10 with the default step 1. More details are given in Sect. 3.8.

"Clump": This parameter aggregates the data of the sonic matrix based on value set here before the application of ACI metrics.

Note that short-term Fourier transform (STFT) with a Hanning window of 1024 points (without overlap) is performed by the SonoScape prior to calculations of ACI. These parameters for STFT are not configurable at this moment.

3.4.2 *Buttons*

There are seven buttons on the main interface (From a to g in Fig. 3.11), including: Three buttons for processing the .wav files:

(a) Perform the calculation for the current selected file in the file list (the "Process" button).
(b) Save the results (the "Save" button).
(c) Move on to the next file (the "Next" button).

Two buttons in the post-ACI process area that perform:

(d) The ecoacoustic events calculation (the "Ecoacoustic Events" button)
(e) The Sonic Signature Dissimilarity (SSD) calculation (the "SSD Analysis" button)

and two "AUTO" buttons:

(f) The "AUTO" button under the Explorer automates the calculations of all files in the folders (and subfolders) listed in the "Folder List".
(g) The "AUTO" button under the "File List" box automates the calculations of all files listed in the "File List".

3.5 Data Management

The SonoScape manages data files in a hierarchical structure. It is a good practice to always save data files in the same folder with an appropriate folder name. SonoScape can operate with different subfolders nested into a principal folder (Fig. 3.12). All results obtained from the SonoScape will be stored in a folder named "Results" within each folder/subfolder. Pictures generated from the SonoScape will be stored in a folder named "Figures" on the same level as "Results folder". Note that the SonoScape will back up automatically the "Results" and "Figures folders before each new run by moving the previous results/figures to folders named by "*Results_time string*" and "*Figures_time string*", respectively. For example, if users perform a new run on April 28, 2021, at 10 in the morning, the two backup folders will be named "Results_20210428100000" and "Figures_20210428100000", respectively.

Figure 3.13 shows a diagram that demonstrates this hierarchical structure with which data are stored by SonoScape after the processing. This configuration depends on the structure of the subfolders submitted to the processing procedure. The example reported here is based on data from 11 recording stations for 4 days of recording. This organization may change according to the goal. For instance, if a multiple scale approach is utilized, or if a range of different energy filters are used, we can have consistent differences in the labels utilized, although the hierarchical structure of folders and data remains identical.

Within the Results folder, there are subfolders, each of which corresponds to one of the audio files and is named with the file's name, too. For instance, the file named 20210323_235400[wav] when processed creates in a folder "Results" a subfolder with the same name of the wav file 20210323_235400[ASCII files].

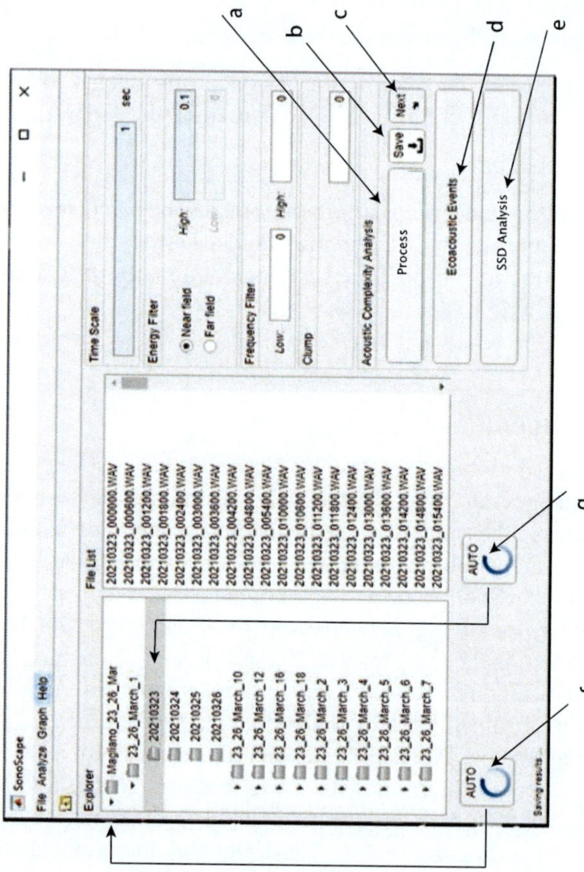

Fig. 3.11 Buttons of the ACI area

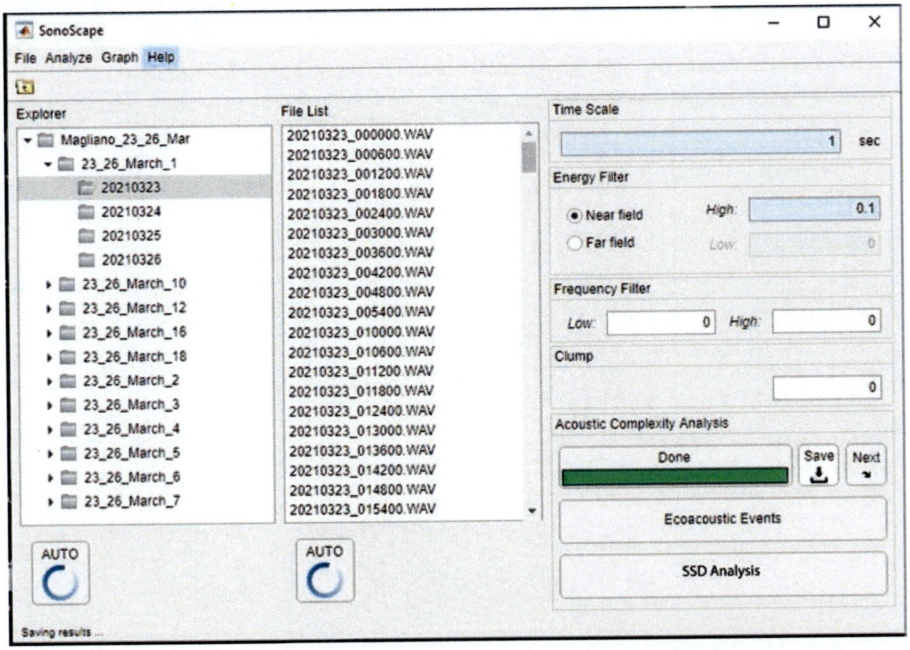

Fig. 3.12 Example of hierarchical organization of files. The Folder Magliano 23–26-Mar contains 11 recording station (RS) subfolders (1, 2, 3, 4, 5, 6, 7, 10, 12, 16, 18). Each RS subfolder contains 4 days subfolder (DS), and each DS contain 240 wav files of 300 second each

Inside the "Results folder", there are also a bunch of Excel files ("ee_*.xslx"; not shown in Fig. 3.13) that summarize the ecoacoustic events (EE) obtained with different combinations of energy filter parameters. For the actual file name, the symbol wildcard * should be the analysis field (near or far), the "Low energy filter", and/or "High energy filter" information. For example, ee_near_0.01.xlsx summarizes results obtained for near field with a "High energy filter" of 0.01. These files have exactly the same layout, and each of them has various numbers of sheets depending on the time scale parameter used to perform such analysis. If the time scale is coincident with the length of a file, we obtain only one sheet for metric.

There are also ASCII files (comma-separated or tab-separated files) that summarize the entropy of the EE ("ee_entropy_*.txt"), the SSD ("ssd_*.txt"), the fractal dimensions of EE, and SSD ("ee_fractal.txt" and "ssd_fractal.txt").

Within each subfolder that is named with the corresponding .wav file name, there are the third-level subfolders named with the specific time scale that the results correspond to. For example, in Fig. 3.13, only one time-scale—1 s—was used to perform the analyses. If other time scales are selected (for instance, four: 1, 2, 3, 4 s), there would thus be four third-level subfolders, each of which stores detailed results calculated under that specific time scale.

The "cache" and "cache_ee" files are encoded binary files for internal use by the SonoScape. They can be read using MATLAB® if one is installed.

"$ACIf_t$ Max.txt" saves the maximum of the $ACIf_t$ matrix which is used for encoding the EE. There is an "ACI" folder within each third-level subfolder, and within

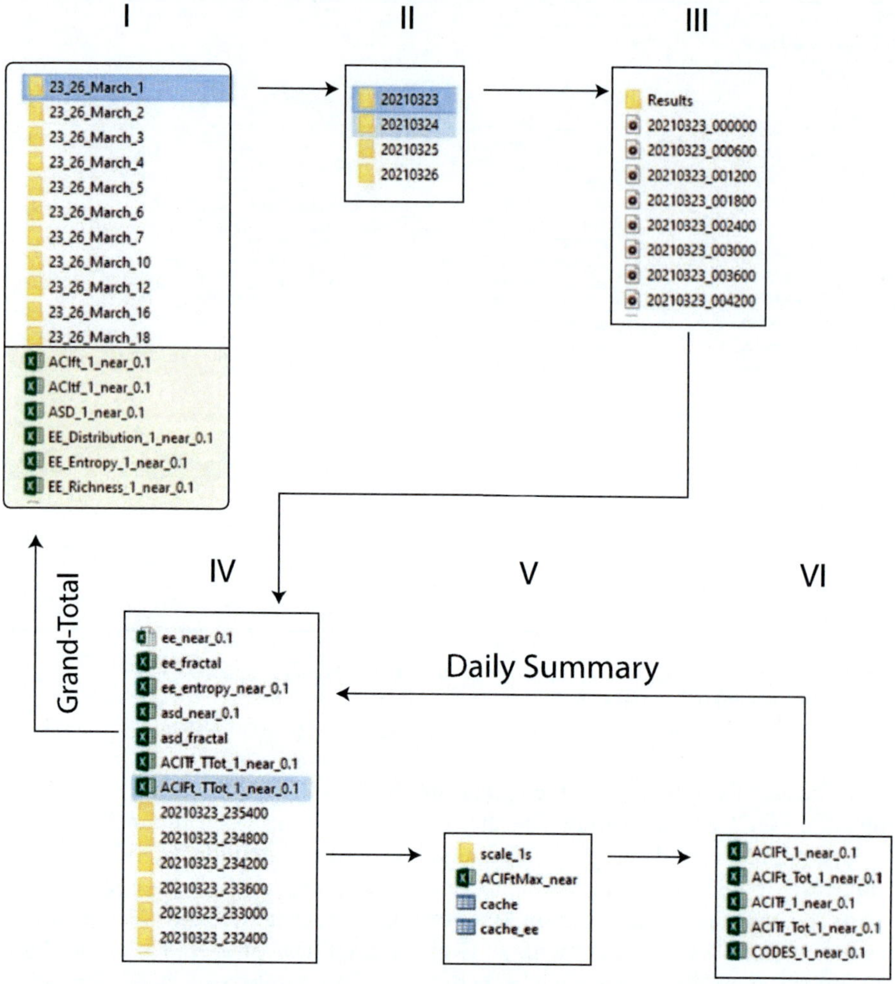

Fig. 3.13 The hierarchical structure of the data managing system used in the SonoScape. This example has six levels. The first level contains 11 recording station data from 23 to 26 March and the Grand-Total files that are printed at the end of the entire process of data processing. The second level visualizes the data of the station #1 (23–26_March_**1).** The third level portrays the results of station #1 and the list of the wav files of this station (240 files each of the duration of 5 min) [In figure only, eight files are shown]. The fourth level portrays the results of each 5′ files and the summary results of the entire day. The fifth level portrays the data of a single wav file named according to the temporal scale utilized, the max of $ACIf_t$ found in this file that will be used for the standardization of Ecoacoustic Events. The sixth level contains the result of the 5′ wav file. Grand total is the result of the aggregation of daily data form level VI that in turn is aggregated at level IV and finally to level I. The trailing strings in the file names generated by the SonoScape (e.g., the files shown on levels I, IV, and VI) are in the format "_time scale_near or far field_energy filters"

this "ACI" folder, there are detailed $ACIt_f$, $ACIf_t$, the sum and evenness of $ACIt_f$, and $ACIf_t$, and the EE results stored in the "ACIFt_timescale_*.txt", "ACITf_timescale_*.txt", "CODES_timescale.txt", and "EE_timescale.txt", respectively.

3.6 Software Configuration

Global settings of the software can be done via the “Options” submenu item under the “File menu” (or through hotkey “Ctrl + O”). The settings window is shown in Fig. 3.14.

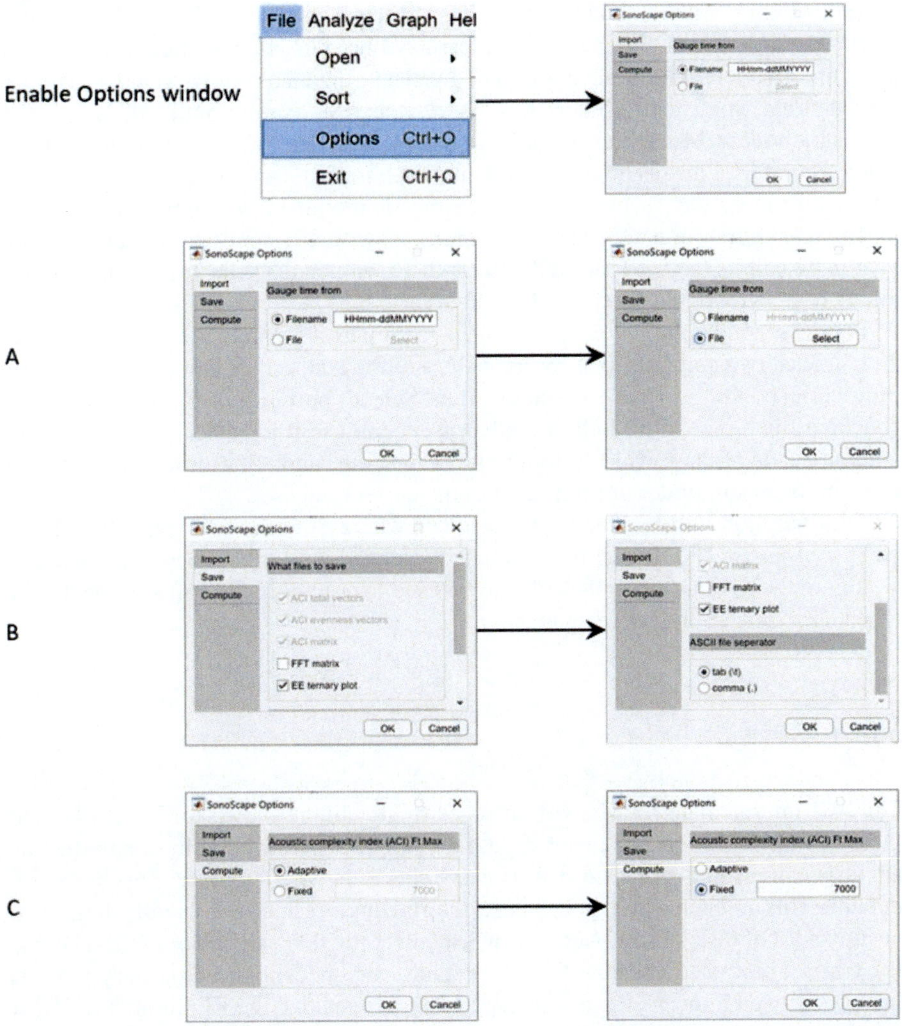

Fig. 3.14 The SonoScape configuration interface. This window will pop up with click on the “Options” submenu under “File” menu. (**A**) The “Options” allow to “Import” files for the sorting routine by using a variable format of file name, or to import from a specific file that contains time for each file. (**B**) Activating the option “Save” is possible to save every FFT matrix according to a tab or common separated format. FFT matrix is important during the choice of the energy filter. It also has the option to export the ternary plot for EE or not. (**C**) With the option “Compute”, it is possible to fix a value of $ACIf_t$ max(in this example $ACIf_t$ max = 7000) to apply to all the wave files processed in a session. If users would like to let the software choose the $ACIf_t$_max value based on the data processed, they can check “Adaptive” here instead of “Fixed”.

3.6.1 Import

The start time of the audio signal imported can be defined through the "Import" tab in the options window. This can be done through either a user-defined string that represents the format of the audio filename (if checking "Filename" radio button), or through another file that specifies the start time for the audio signal under analysis (if checking "File" radio button).

For the first option "Filename", wildcard symbol * is allowed, but each * represents one character in the file name. The format string also follows standard syntax for date/time. Specifically, letter "Y/y" is for year; "M" is for month; "d" is for day; "H" is for hour in 24-h style; "h" is for hour in 12-h style; "m" is for minute; and "s" is for second. HH means hour in two digits, 24-h format; mm means minutes in two digits; dd means day in two digits; MM means month in two digits; and YYYY means year in four digits. Some other common symbolic identifiers include YY for year in two digits (e.g., 20 for 2020, 19 for 2019, etc.) and MMM for month in three letters (e.g., JAN for January, FEB for February, etc.).

For the second option "File", users need to prepare a separate file in Excel format with at least two columns, one for the names of the audio files, and one for the start time corresponding to each of them. The "Select" button will be activated after checking the "File" radio button so that users can use it to select such a file. The column heads of the Excel file should be "Filename" and "Starttime", respectively (this option is still under development in the current release).

Configuration will be stored in the folder where SonoScape is installed via an encrypted binary file. It will be automatedly loaded next time when users want to configure the options again. It will also be used to configure the software next time when users open SonoScape again.

3.6.2 Save

The ACI intermediate results will be saved by default as shown in Fig. 3.14a, and these options cannot be changed by users. The intermediate results for frequency analysis can also be stored in disk if users check the "FFT matrix" checkbox. By default. This is unchecked since these intermediate results are usually large and occupy a lot of disk spaces. Another intermediate file that can be configured here is the ternary plot used to show EE. By default, the checkbox "EE ternary plot" is unchecked which means the SonoScape will not visualize the EE automatically. If checking this checkbox, the SonoScape will show a ternary plot after running all files within a subfolder (that means all files listed in the File List) that visualize the EE, and export the plot to a .pdf file in the "Figures" folder.

The ASCII file separator setting item is used to specify which symbol will be used for all ASCII files generated by this software. There are two options: tab (\t; the default option) or common (,).

3.6.3 Compute

"Compute" button (Fig. 3.14c) allows to select the max value of ACI_t in an entire collection of data to standardize the $ACIf_t$ for the calculus of Ecoacoustic Events. If the adaptive button is selected, the $ACIf_{t\ max}$ is selected inside each folder.

3.7 Open

The SonoScape is designed in a way that can process multiple files as a batch. The "Folder" sub-submenu item under "Open" submenu item (see Fig. 3.8) is to locate a folder that stores all the audio files to be analyzed. After that, the file names will be listed in the file list area on the main interface. This can also be done by using the hotkey combination "Ctrl+F".

3.8 ACI Analysis

In the ACI area on the main interface (see Fig. 3.9), there are edit boxes to type in related parameters as discussed in Chapter 2. Specifically, the "Time scale" edit box accepts input for defining the time scales (in unit of second) that are used to perform ACI analysis. This edit box allows expressions with simple syntaxes (Fig. 3.15). Some hints information will be displayed if users hover over the edit box.

For example, if users want to perform the analysis under time scales from 1 s all through to 10 s, they can use either [1, 2, 3, 4, 5, 6, 7, 8, 9, 10] in this edit box, or simply 1:10 which means to generate a vector from 1 to 10 with the default step 1.

If users would instead like to perform the analysis under odd numbers from 1 to 10 s, they can use 1:2:10 which means to generate a vector from 1 to 10 with step 2 (this is equivalent to [1, 3, 5, 7, 9]). If we use 1:2:20, the scale utilized will be 1, 3, 5, 7, 9, 11, 13, 15, 17, 19. In the case of scale setting using x:y:z, x and z give the lower and upper limit, whereas y gives the step. In mathematics, the second number equals $x + y$; the third number equals $x + 2y$; and thus the i-th number equals $x + (i-1)y$.

Similar principles apply to the "Low energy filter" and "High energy filter" edit boxes as well. The remained edit boxes accept only a single number (or a scalar) as their inputs.

After setting up these input parameters, users can push the "Process" button to perform the analysis on the current selected file highlighted in the "File list". By default, the first file will be selected and highlighted after importing. There will be a progress bar displayed on the "Process" button to show the analysis progression,

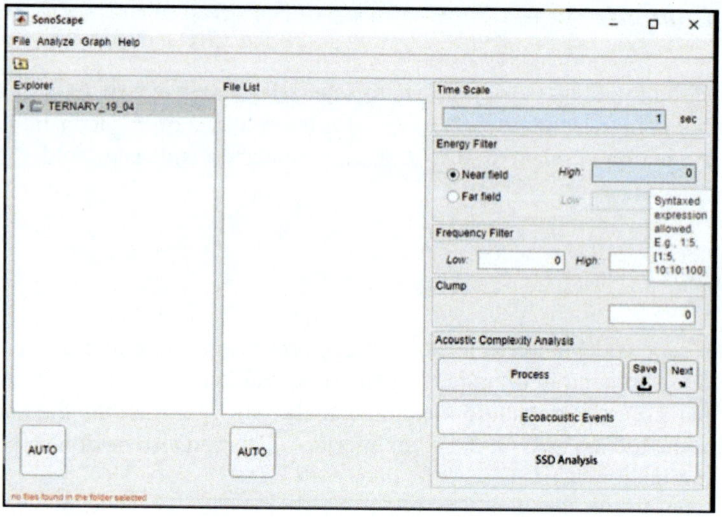

Fig. 3.15 ACI analysis area. Shown the hints information if hovering over the "Time scale" edit box

and the calculation status will be updated in real-time on the status bar area in the lower left corner (Fig. 3.16).

After this step, the calculated results will only be saved in the memory (the random-access memory, RAM) of the computer. Users need to click the "Save" button to write them to disk in a location that is managed by the SonoScape. After that, users can either click "Next" button to move to the subsequent file below the current one or select another file on the list by left clicking on the specific file and push the "Process" and "Save" again to perform the analysis and save results. Alternatively, if users would like to apply such analyses to all files listed in the "File list", they can simply push the "AUTO" button after importing. The software will automatically perform the analyses and save results for all files, or all files below the current selected file (included) if users push "AUTO" button after select a specific file on the "File list". During this process, the "AUTO" button will display a "busy" icon, while the progression will be shown on the "Process" button with a progress bar (see Fig. 3.16) and updated in real-time in the status bar area.

3.9 Post-ACI Process: The Ecoacoustic Events and the SSD Calculations

The ecoacoustic events calculation can be done by clicking the "Ecoacoustic Events" button in the Post-ACI process area. This calculation is automatically performed throughout the files, and actually this button will only respond after ACI calculations are done for all files in the "File list", since the encoding of ecoacoustic

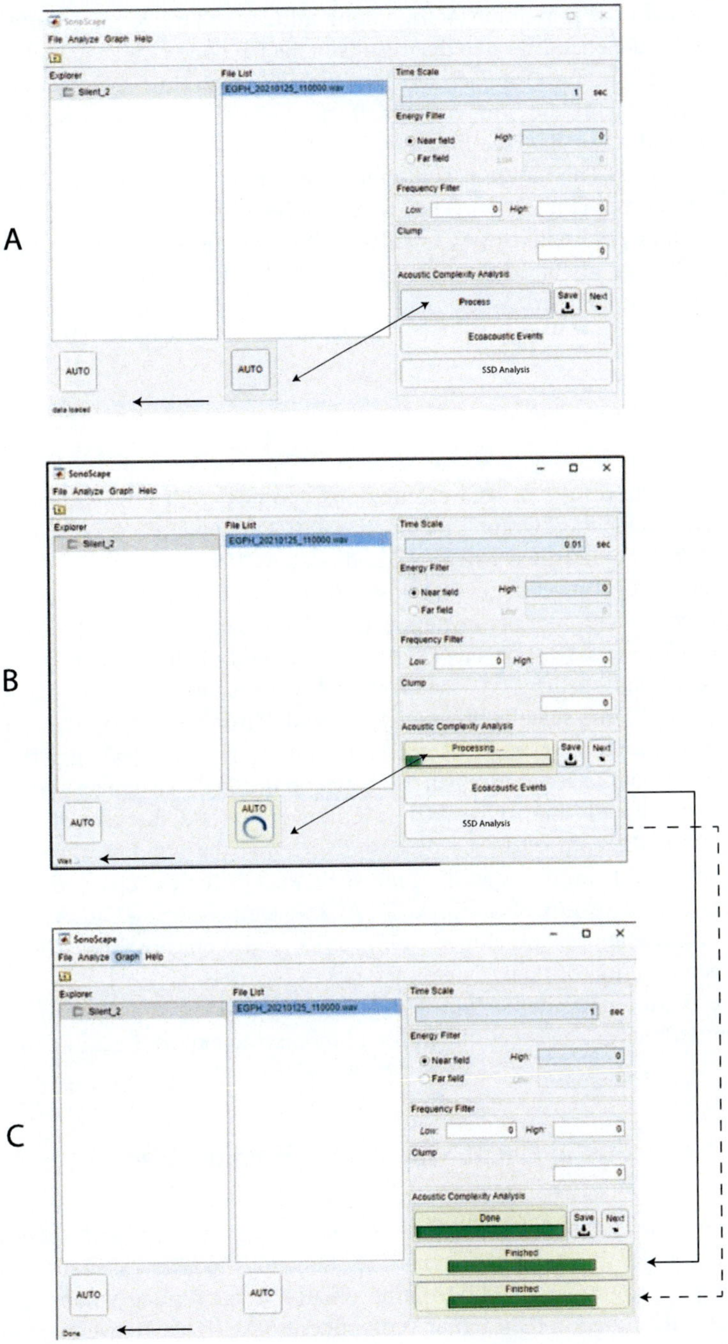

Fig. 3.16 Performing ACI analysis. Shown the progress bar and real-time status information from (**A**) to (**C**)

events requires the maximum $ACIf_t$ among all to standardize the results. This process includes the encoding step that results in the EE codes, and calculating step that calculates the frequency of each EE pattern under each time scale, and the entropy results of EE. Again, the progression and status of the calculations are displayed on the button and status bar (see Fig. 3.16).

SSD analysis. The button is right below the “Ecoacoustic Events” button in the lower right corner (Fig. 3.16). The “SSD Analysis” is only activated after all files have been analyzed with the ACI analysis. It is also programmed to be fully automated, and users just need to click it once which will run the analysis through all files listed in the “File list”.

3.10 ACI Visualizer

To visualize the results of ACI calculations and post-ACI process, there is a submenu item under the “Graph” menu—the “ACI Visualizer” that is programmed. This item will be activated when ACI and post-ACI calculations are done. An ACI Visualizer configuration panel (see Fig. 3.17) will be first popped out for users to select the combination of parameters with which the corresponding results are to be visualized. Figure 3.18 shows an example of the ACI visualizer window.

Before activating the ACI visualizer, a file from the “File list” panel must be selected. The “Time scale”, “Low energy”, and “High energy” in the “ACI visualizer” configuration panel vary based on the input users defined when performing the ACI calculations. There are six panels shown on the “ACI visualizer” window. They visualize specifically the $ACIf_t$, $ACIt_f$, $ACIf_t$ evenness, $ACIt_f$ evenness and entropy and SSD results from post-ACI process. Note that the window shows each time results obtained from one specific time scale except for entropy and SSD results. For example, the results of $ACIf_t$, $ACIt_f$, $ACIf_t$ evenness, and $ACIt_f$ evenness calculated on time scale 1 s are shown in Fig. 3.18, while entropy and SSD results are shown as a function of time scale on the last two panels.

There is one menu item “File” for this window and a submenu item “Export” programmed there. The function of this submenu is to export the figure to hard disk.

3.11 Ecoacoustic Event Visualizer (Ternary Plot)

To visualize the Ecoacoustic Events distribution on a ternary plot, there is a submenu item under the “Graph” menu—the “Ecoacoustic Event Visualizer” (Ternary Plot) that is programmed. This routine creates ternary plots where Ecoacoustic Events are illustrated in their ternary components ($ACIf_t$, $ACIf_t$ evenness, and $ACIt_f$ evenness values are used as barycentric coordinates) (for more details about the history of the ternary diagrams, see Howarth 1996). The size and color of the circles indicate the frequency of occurrence of each code. The scale of representation is

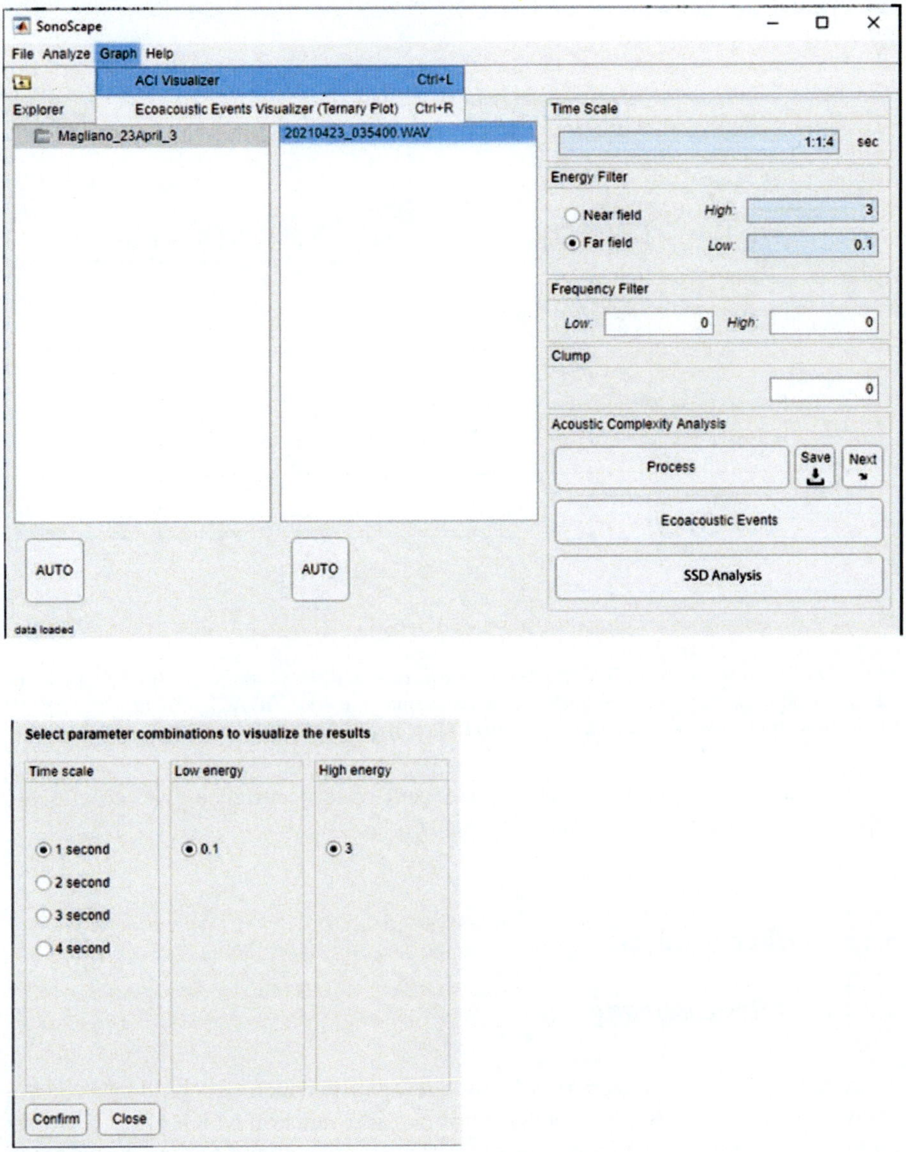

Fig. 3.17 ACI visualizer routine allows the representation of $ACIf_t$ and $ACIt_f$ in a 3-D diagram, $ACIf_t$ evenness, $ACIt_f$ evenness, entropy and SSD results as line graphs

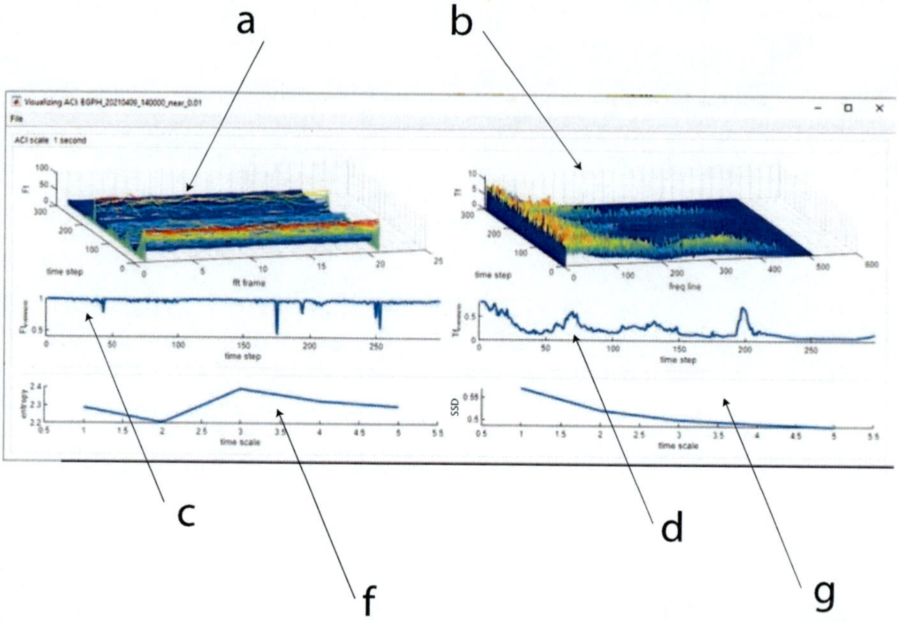

Fig. 3.18 The SonoScape ACI visualizer window. (**a**) $ACIf_t$ (y = 300 s; x = FFT frame). (**b**) $ACIt_f$ (y = 300 s; x = 512 spectral lines). (**c**) $ACIf_t$ evenness in 300″. (**d**) $ACIt_f$ evenness in 300″. (**f**) EE-entropy at five temporal scales (1–5 s). (**g**) SSD at five temporal scales (1–5 s)

logarithmic in order to exalt the less frequent code occurrence. The color scale depends on the relative abundance of code (Fig. 3.19).

3.12 Other Analysis

3.12.1 Files Rearrangement

There are occasions that users may want to reorganize data files should there be too many in a folder. The SonoScape has built in a sort function which offers file rearrangement by the proprietary dates of files. This routine allows to sort in a folder in which wav files are originally sorted by time and not by day (f.i., SET^{tm} recorders [www.Lunilettronik.it] write files in this matter) creating separate folders named according to the day. This function can be useful to separate files per days in folders that contain more days, avoiding in this way to select and separate by hand each list of files. From the menu "File", users can select the submenu "Sort", and then select the sub-submenu "By Day". A window will appear asking users to define a directory that contains data files to be sorted. The rearrangement will be running in background after defining the folder and clicking "Select Folder".

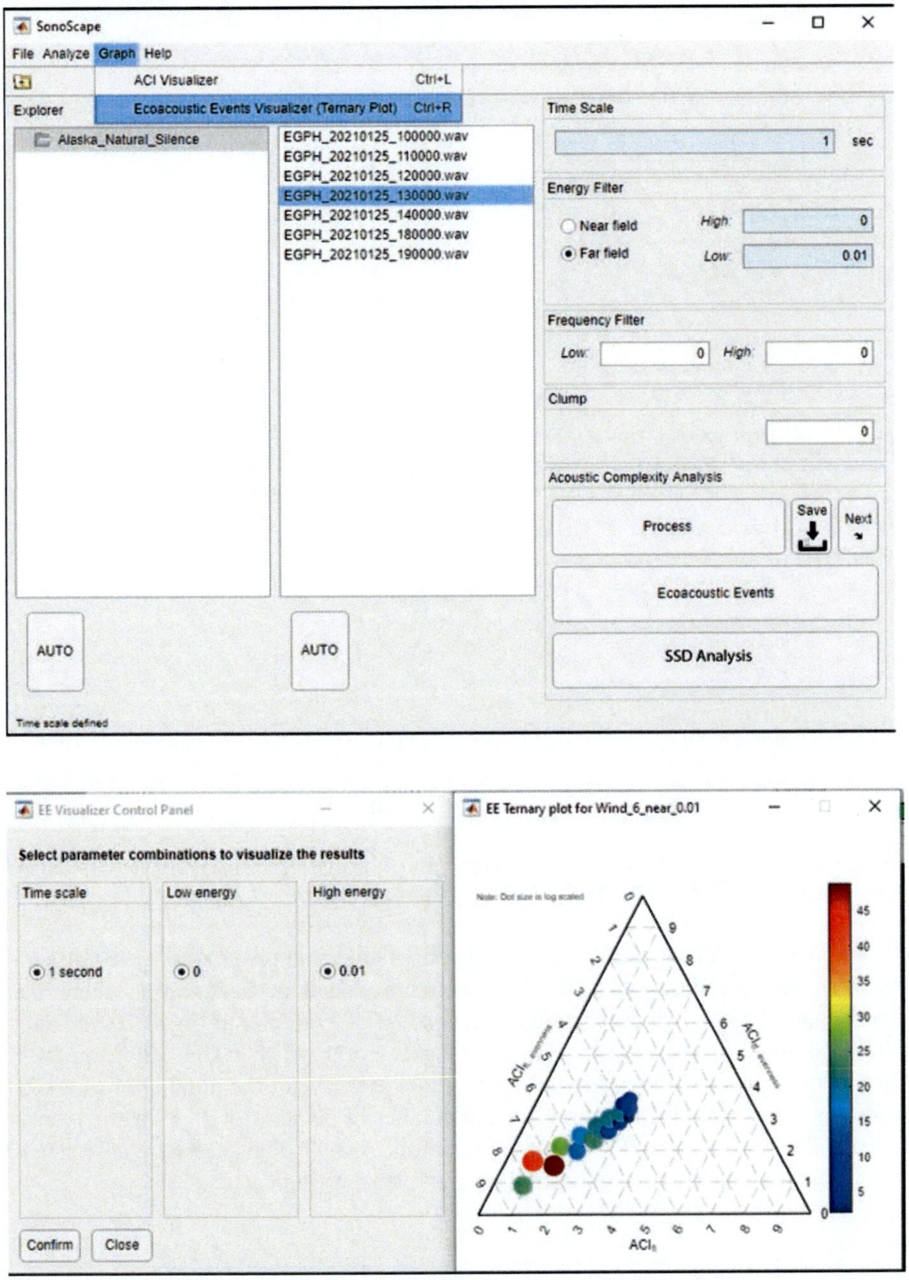

Fig. 3.19 "Ecoacoustic event visualizer". The distribution of ecoacoustic events is portrayed according to a ternary plot where a given plotted circle represents the relative proportion (barycentric coordinates) of three indices ($ACIf_t$, $ACIf_t$ evenness and $ACIt_f$ evenness)

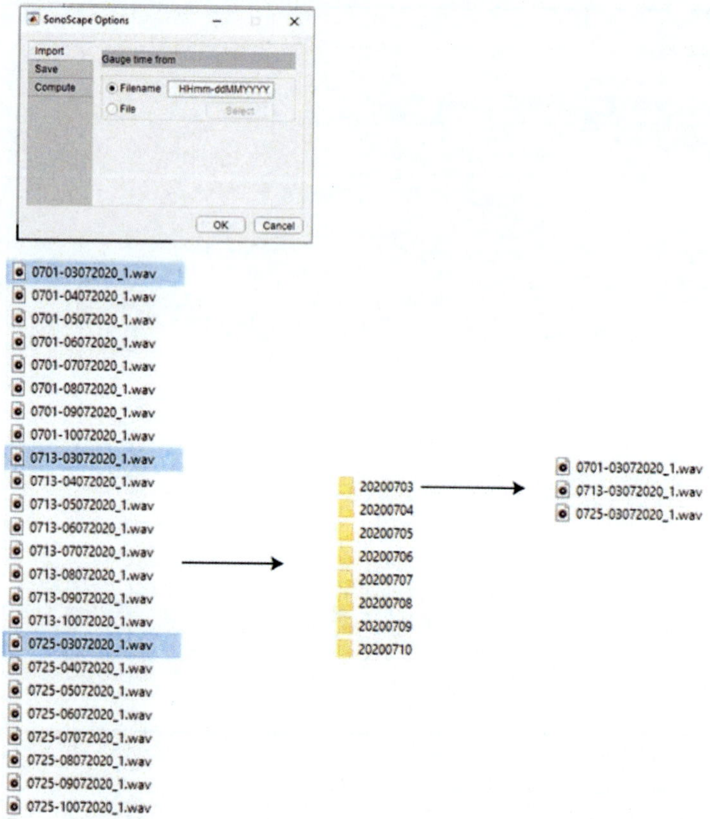

Fig. 3.20 Example of application of the routine "Sort" to twenty-four files originally sorted by hours and minutes. The routine "Sort" creates in this case 8 folders each containing three files

Note that the sorting function depends on whether the proprietary dates of data files can be parsed or not. This has been introduced in Sect. 3.6.1 where the SonoScape "Option" configuration is discussed. For example, in Fig. 3.20, the data files are named in the format "HHmm-ddMMyyyy*******" [HH = hour in 24-h format, mm = minute, dd = day, MM = month, yyyy = year; the number of asterisks indicates some other information like the number of device (f.i., "_1" is the identifier of a SET[tm]) plus four **** for the extension ".wav"]. The letters are case sensitive so please consider using the appropriate format as indicated.

3.13 About Information

Under the "Help" menu, there is one submenu item, the "About" submenu, which will open the about information of the SonoScape software application (Fig. 3.21).

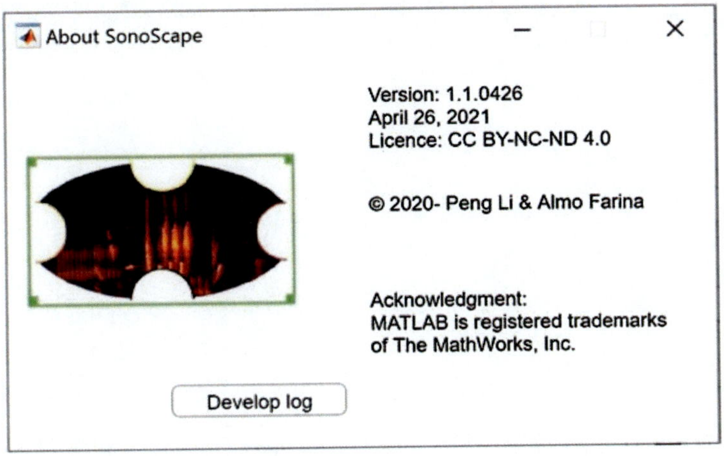

Fig. 3.21 About SonoScape

References

Farina, A., & Salutari, P. (2016). Applying the ecoacoustic event detection and identification (EEDI) model to the analysis of acoustic complexity. *Journal of Mediterranean Ecology, 14*(14), 13–42.

Howarth, R. J. (1996). Sources for a history of the ternary diagram. *BHJS, 29*, 337–356.

Li, P., & Farina, A. (2021). SonoScape. Retrieve from https://github.com/pliphd/SonoScape

Sueur, J., Aubin, T., Simonis, C., Lellouch, L., Brown, E. C., Depraetere, M., Desjonqueres, C., Fabianek, F., Gasc, A., Kasten, E., LaZerte, S., Lees, J., Marchal, J., Mikulec, A., Pavoine, S., Pinaud, D., Stotz, A., Villanueva-Rivera, L. J., Ross, Z., Witthoft, C. G., Zhivomirov, H. & Libsndfile, S. (2020). *Package 'seewave'*.

Villanueva-Rivera, L. J., & Pijanowski, B. C. (2018). Soundecology (Version 1.3.3) [R: A language and environment for statistical computing]. Retrieved from https://CRAN.R-project.org/package=soundecology.p

Chapter 4
Exercises

Abstract SonoScape is a powerful software devoted to the computation of sonic files by the application of the Acoustic Complexity Indices. The illustration of detailed exercises regards primarily:

1. How to set the Energy filters reducing the discretion in the selection of thresholds, a necessary step to correctly apply the ACI metrics and to obtain the sonic and acoustic signatures.
2. Working with the energy filter in order to exclude every signal that does not belong to environmental sonic sources and that mask biophonic signals.
3. Working with the sonic signature, discovering the use of this semiotic object obtained by using $ACIt_f$.
4. Working with near/far sonic field. This exercise introduces to the delimitation of far and near fields, a necessary passage to couple sonic to landscape data.
5. Working with the Sonic signature across landscape.
6. Working with the temporal scale. The aggregation of the ACI metrics at different temporal scales allows to evaluate the effect of temporal aggregation.
7. Working with the fractal dimension of the Ecoacoustic Events at dawn. In this exercise, the fractal dimension of ecoacoustic events has been calculated.
8. Graphing Ecoacoustics Events. Plotting the ecoacoustics event codes, composed of three digits into a ternary plot return a representation of the spatial distribution of the ecoacoustics events. This design is useful to make confrontation between different files/location/time, etc.

Keywords Demonstration of *SonoScape* performances · Energy filters · Sonic signature · Ecoacoustic events · Fractal dimension of ecoacoustic events · Ternary plots

A. Farina, P. Li, *Methods in Ecoacoustics*, Frontiers in Ecoacoustics 1,
https://doi.org/10.1007/978-3-030-82177-7_4

4.1 Introduction

This chapter does not cover every potentialities of the SonoScape software but aims to describe the major utilities in an attempt to tie the sonic semantic with the ecological agents.

The exercises are presented mainly with figures that are the carbon copy of the SonoScape output.

The first exercises are focusing on the setting of energy filter, an obligated passage to an appropriate use of ACI metrics. Then some exercises related to the sonic signature and the ecoacoustics events are presented. Finally examples of a Sonic Information System and of the ternary plot applied to ecoacoustics events are presented.

4.2 How to Set the Threshold of the Energy Filter

To reduce the arbitrariness in the choice of the energy filter values necessary to eliminate every nonenvironmental signals before processing ACIs in a collection of sonic files, you can utilize a direct inspection of a file that has no environmental signals. For instance, a file recorded at night (may be in winter time), when there are no sounds at all (no wind, rain, or biophonies and absence of anthropophonies or technophonies), is the best candidate.

Here the entire procedure:

1. Before running SonoScape for the ACI computation, select the FFT matrix box in the SonoScape "Options" (Fig. 4.1) and then tap **"save tab"**.
2. Set the temporal "scale" to compute a unique FFT matrix of your file. The temporal scale must have the same duration of the sonic file (in our case 300 second having used a file of 5 minute). If the temporal scale is equal to the length of the files, the program returns a single FFT matrix (i.e., performing the FFT on the entire collection of data in the file). Otherwise, it returns a number of matrices

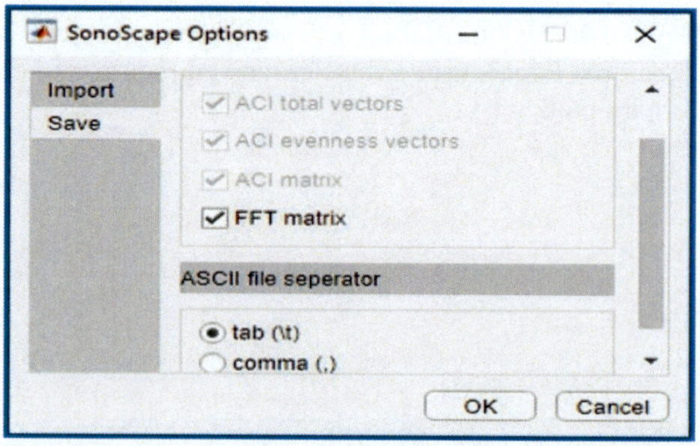

Fig. 4.1 The interface to save the FFT matrix

according to the temporal intervals of analysis. For instance, if we select 1 second as time scale, 300 FFT matrices will be produced (i.e., performing 300 FFTs each on a 1 second segment of the recording). Note, what the program actually does is the short-term Fourier transform (STFT) using Hanning window of length 1024 without overlap. Therefore, one dimension of the FFT matrix is time, and the other dimension is frequency. The values are the amplitude of the Fourier spectrum that represent the energy of the recording corresponding to the specific time and frequency defined by its horizontal and vertical axes.

3. Select a file that has not structured signals based on a visual inspection of its spectrogram.
4. To improve the visibility of the spectrogram and to be sure of the choice, increase the intensity of the spectrogram using an acoustic editor (Fig. 4.2).
5. If this file has no signals, it represents the ideal baseline to fix appropriately the energy filter to be applied to the entire collection of data. Then, search for the max value of the FFT present in the matrix. In our case, a matrix with a dimension 512 × 14,062 is composed by 7,199,744 values. Note, the number of columns should be equal to the floor integer of data length in second × sampling frequency in Hz/1024; the number of rows equals to half the window length (one-sided spectrum).
6. Select the elements in the FFT matrix according to categories of intensity filters that you believe valuable. For instance: 0.001, 0.05, 0.1, 0.5, and 1. The numerosity of each category is presented in Fig. 4.3.

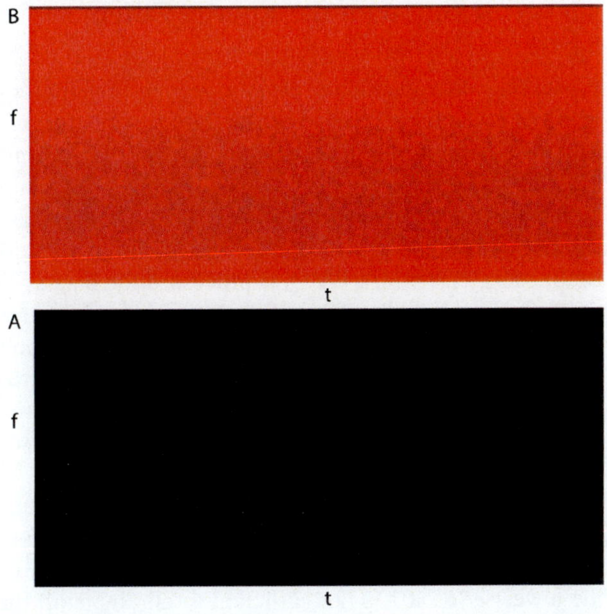

Fig. 4.2 (**A**) Spectrogram of a file of 300 s recorded at 00:18 on March 10, 2021 (at Magliano study site [44°14′, 10°03′ E]). (**B**) Enhanced effect after the application of 40 dB. f = frequency, t = time

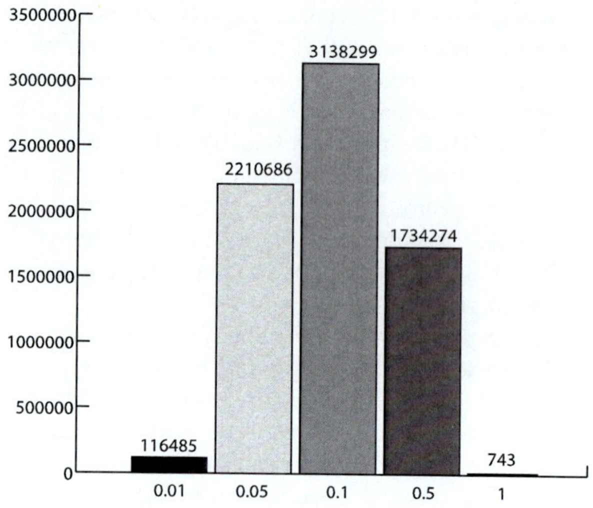

Fig. 4.3 Distribution of five energy filters (0.01, 0.05, 0.1, 0.5, and 1) of a sonic matrix of 512 × 14,062 elements after the application of an FFT to a wav file

7. The results establish that if an energy filter equaling to one is selected, the entire collection of nonenvironmental signals should have been eliminated from this FFT.
8. Finally process the entire collection of data using the energy filter that ignores all the data present in the FFT matrix. If the filter is appropriate, you should have all the metrics equal to 0.

4.3 Working with the Energy Filter

Processing data excluding every nonenvironmental signals or signals extremely feeble is essential to have understandable results from ACI metrics. We present two examples from different ecoregions, the first from Madagascar and a second from Mediterranean region.

In this first example (Fig. 4.4) from a Madagascar data set (January 5, 2013) (Courtesy of Anne Axel), our goal is to demonstrate the importance to utilize an adapt energy filter (*ef*). We have processed the data at 1 second of temporal resolution with four energy filters: 0.01, 0.05, 0.1, and 0.5, respectively. As expected, the lowest energy filter (*ef* = 0.01) is not sufficient to eliminate all the signals that mask the real trend of such soundscape. From our experience (AF), it seems that the best *ef* is equal to 0.1. The inspection of the sonic signatures (frequential and temporal) is the best procedure to do this evaluation, although also the other ACI metrics show a clear sensitivity to the adopted filter.

The second example is from a Mediterranean data set (Magliano (44°14′ N, 10°03′ E), March 10, 2021) (Fig. 4.5). Data were processed at 1 second of temporal resolution with five different *ef*: 0.01, 0.1, 0.25, 0.5, and 1, respectively. As expected, the lowest *ef* = 0.01 is not sufficient to eliminate all the nonenvironmental signals

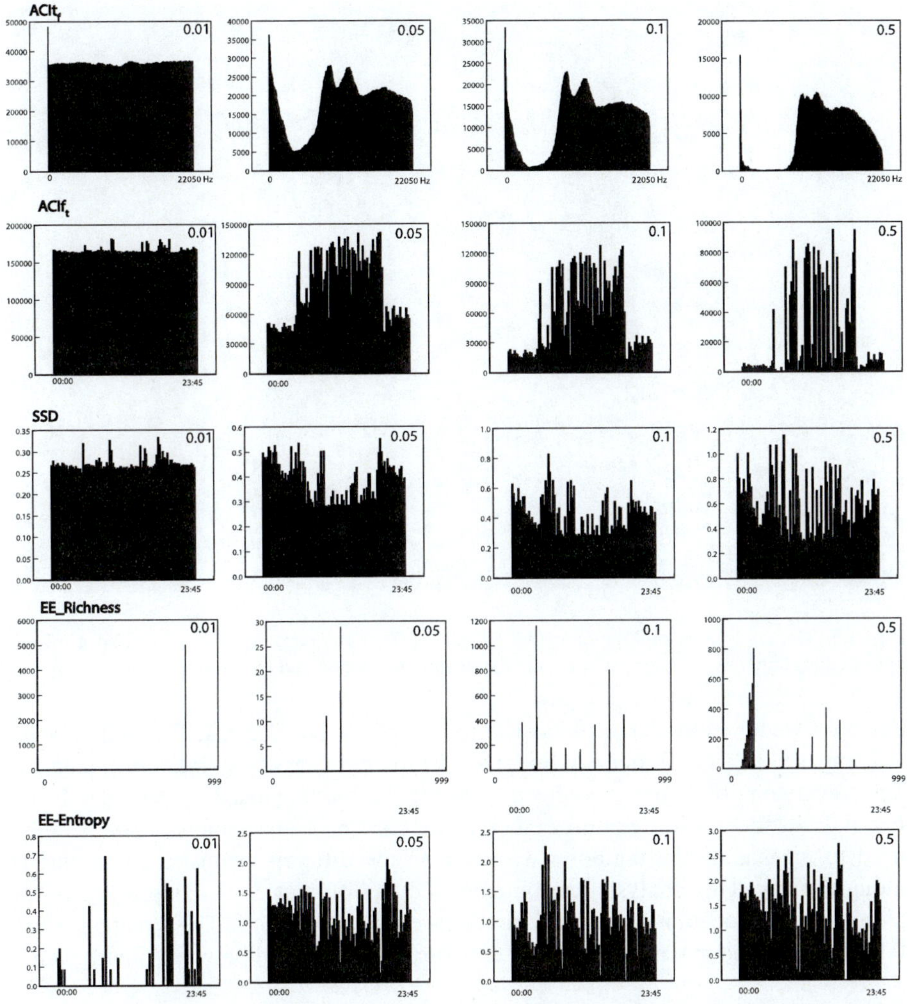

Fig. 4.4 It is evident that the exclusion of values = < 0.01 is not sufficient to "clean up" the data and that the higher values of energy filter (ef = 0.5) exclude also environmental signals

that mask the real trend of this sonoscape. Like for the example of Fig. 4.6, it seems that the best *ef* is the one with value 0.1. The sonic signature and the EE richness seem the best candidates in the choice of the energy filter (Fig. 4.6).

4.4 Working with the Sonic Signature

In this example, we present the potentiality of $ACIt_f$ to describe the sonic signature at hourly scale. Sonic signature is an important descriptor of a sonotope and can be utilized to make comparisons between daily periods or between locations. In this

Fig. 4.5 The study area of Magliano (44°14′ N, 10°03′ E) is represented by a "coltura mista" system dominated by olive groves in some extension abandoned and by dense oak woodlands

example, we have utilized the data of Magliano of March 10, 2021. The $ACIt_f$ matrix is composed of 512 (frequency lines) × 240 five-min sessions (collecting data for 5 min every 6 min). Values have been aggregated according to a 24-h calendar. From Fig. 4.7, distinct sonic signatures emerge along the day. The sonic signature can be used to investigate the temporal dynamic on the different acoustic communities along the day. This analysis can be extended to different days along a season or along different locations. The night hours are easily distinguished by the other hours (GMT time). The maximum of $ACIt_f$ complexity is observed between 5 and 7 a.m.

4.5 Working with Near/Far Acoustic Field

In this exercise, we set the energy filter according to seven combinations between near and far buttons. The entire set of simulations may be extended at any temporal scale.

With the combination of far and near buttons, it is possible to select sounds with energy higher than a threshold selected according our requirement, or based on an objective field mensuration or to select an energy belt in which to apply the ACI metrics.

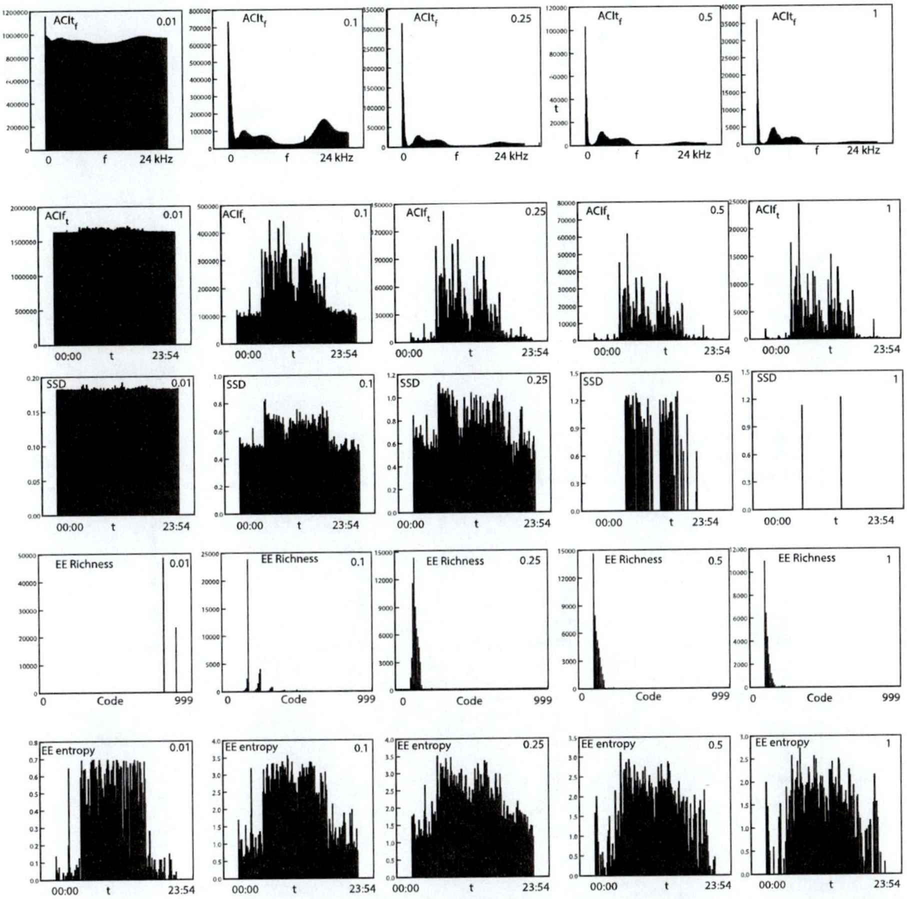

Fig. 4.6 Confrontation of ACI metrics when applied by different energy filters (0.01, 0.1, 0.25, 0.5, 1) at the Magliano study site (44°14′ N, 10°03′ E)

For this exercise, we have utilized a 3 min and 32 s wav file sampled at 48 kHz and containing, after an aural inspection, dominant songs of thrushes (*Turdus philomelos*), at least one individual far and one closer, in addition an alarm call of a chaffinch (*Fringilla coelebs*), blackbird (*Turdus merula*), European turtle (*Streptopelia turtur*), treecreeper (*Certhia brachydactyla*), and a baking dog. We present seven combinations to explain how to apply the filter (Fig. 4.8).

Now are presented the different combinations between near and far bottom:

Combination #1

In this case, we select the button *far = 0*. No filter is applied to the sonic matrix (Fig. 4.9).

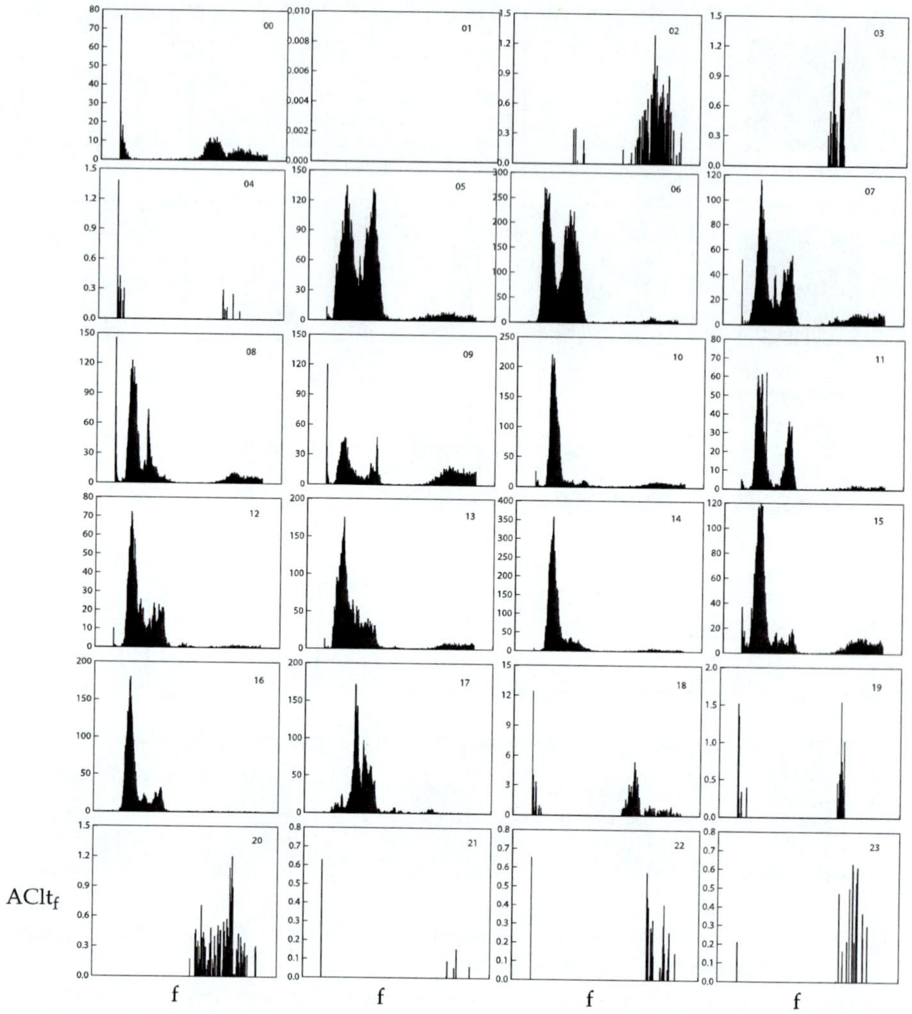

Fig. 4.7 Shape of the sonic signature at every hour of the day (Magliano study site) (44°14′ N, 10°03′ E)

Combination # 2

In this case, we checked "near field" and set "High" (energy filter) = 2. The ACI is calculated for every values of the sonic matrix > of a threshold fixed at 2 (Fig. 4.10).

Combination #3

In this case, we select the button *near* = *2:2:6*. The first number 2 is the value utilized for the first calculation. The second value represents the increment to be

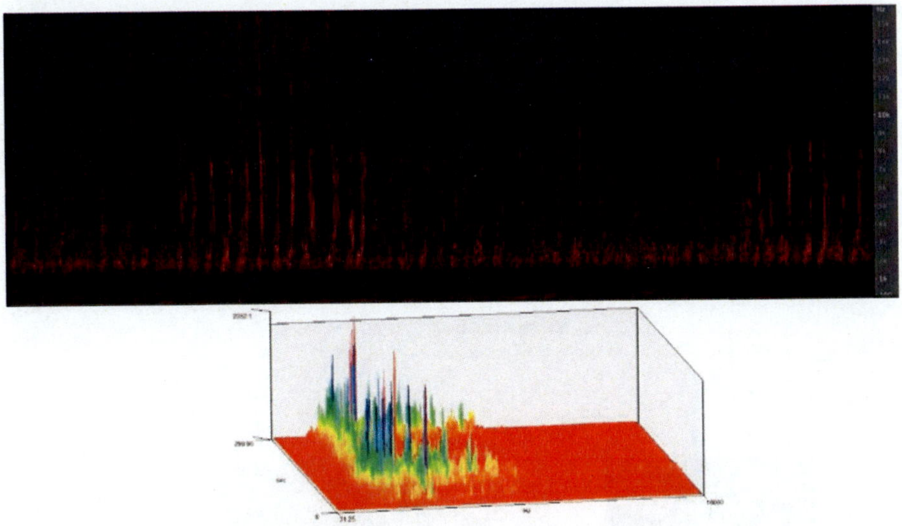

Fig. 4.8 Spectrogram utilized in the application of the energy filter according to the different combinations offered by the software and its 3D representation (Magliano (44°14′ N, 10°03′ E), April 23, 2021, at 05:54 a.m

applied to the first value. In this case is 4 after the first calculation. The last value is the last threshold from which to apply ACI metrics (Fig. 4.11).

Combination #4
In this case, we select the button *far* = 1:2 and the button *near* = 5. We apply ACI metric to belts that are ranging from 1 to 5, and from 2 to 5 (Fig. 4.12).

Combination #5
In this case, we select the button *far* = 1:1:3 and the button *near* = 5. We apply ACI metric to belts that are ranging from 1 to 5, 2 to 5, and 3 to 5. The second value (1) of the far button is the increment to be applied in the range 1–3. The value *near* = 5 represents the max threshold of the belt (Figs. 4.13 and 4.14).

Combination #6
In this case, we select the button *far* = 1:1:3 and the button *near* = 4:5. We apply ACI metric to belts that are ranging from 1 to 4, 1 to 5, 2 to 4, 2 to 5, 3 to 4, and 3 to 5. The second value (1) of the far button is the increment to be applied in the range 1–3. The values 4–5 of the near button represent the two max thresholds of the belt.

In this case, we select the button *far* = 1:1:3 and the button *near* = 4:1:6 and we apply ACI metric to belts that are ranging from 1 to 4, 1 to 5, 1 to 6 then 2 to 4, 2 to 5, 2–6, and finally 3 to 4, 3 to 5, and 3 to 6. The second value (1) of the far button is the increment to be applied in the range 1–3. The second value of the near button represents the increment to apply between 4 and 6. Finally, we obtain 9 belts. This combination may be extended to many other intervals and increments (Fig. 4.15).

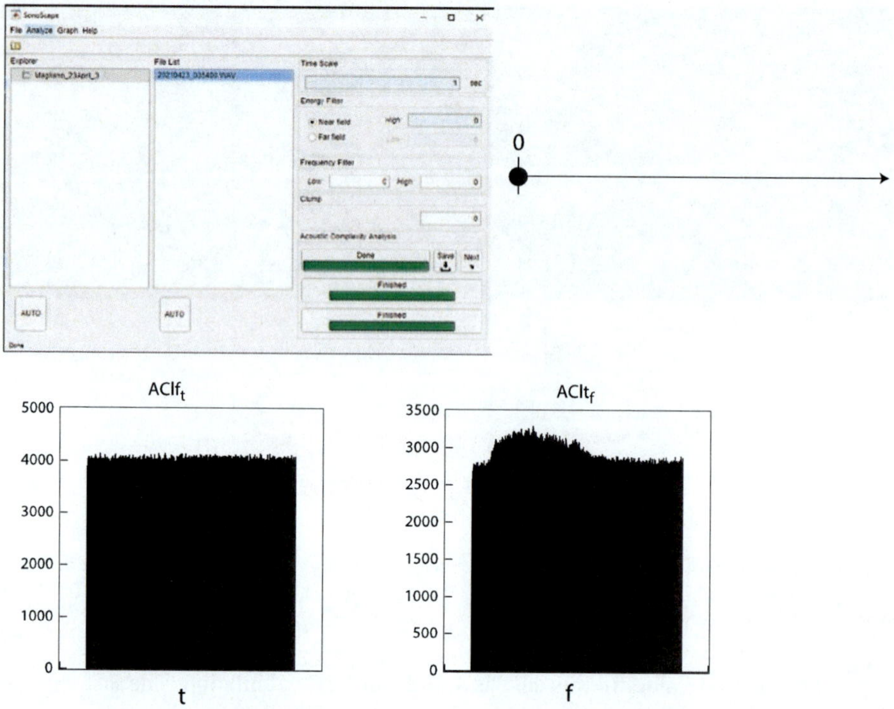

Fig. 4.9 Temporal and spectral sonic signature without the application of an energy filter. (Magliano (44°14′ N, 10°03′ E), April 23, 2021, at 05:54 a.m., recording station #16)

4.6 Working with the Sonic Signature Across Landscape

With this exercise, we intend to show how to manage a set of 11 recorders (AudioMoth™) deployed for 4 days in the Magliano study area (44°14′ N, 10°03′ E).

Every recorder has been set to record 300 s with a pause of 60 s at a sampling rate of 32 kHz. We collected 240 files a day for consecutive 4 days using 11 recorders, resulting thus in 10,560 files, each of 300 s in duration. It needs a memory of 18.3 MB for every 300 s, or in total, 193.248 GB of hard disk space for all of them.

The time requested for the elaboration of the files was 34260 s, or in other words, 9 h and 31 min. This can be done automatically with SonoScape after the sorting of files each day from the SD card in which the files of each recorder are posed into a simple list.

Using the procedure of sorting, it is possible to create as many folders as it requires (in this case four folders since we have 4 days of data) for each recorder and aggregate all the folders corresponding to each day of each recorder in a higher level folder labeled using the interval of recording days (in this case from March 23 to 26, 2021). Finally, all the recording folders are aggregated into a folder that reports the name of the localities and the interval of time considered.

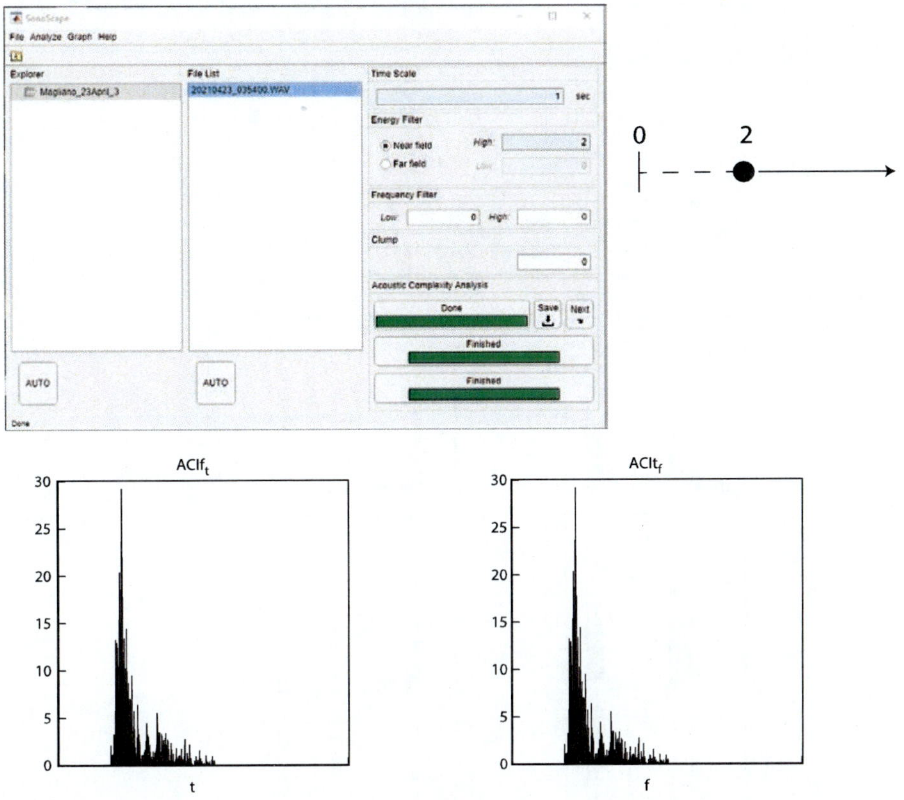

Fig. 4.10 Temporal and spectral sonic signature with the application of an energy filter = 2. (Magliano (44°14′ N, 10°03′ E), April 23, 2021, at 05:54 a.m., recording station #16)

In Fig. 4.16 is reproduced the study area with the recording stations inside hexagons and the screen shot of the SonoScape program. In Fig. 4.17 is represented the complete sequence of the nested folders. Data were processed at the temporal scale of 1s, with an energy filter set at 0.1.

In Fig. 4.18 are illustrated the sonic signatures of each day for the 11 sampling stations.

The sonic signatures obtained using the ACIt$_f$ metrics are reported in Fig. 4.18. The first 10 spectral lines are not represented to better highlight the biophonic components.

Despite a reduced temporal window of 4 days (23, 24, 25, and 26), the difference between the recording stations and at the same time the high similarity between the days in each station are evident.

In Fig. 4.19 is represented ACIf$_t$ that is a measure of the temporal variation of sonic information (temporal sonic signature). In this case, ACIf$_t$ is represented along the entire day.

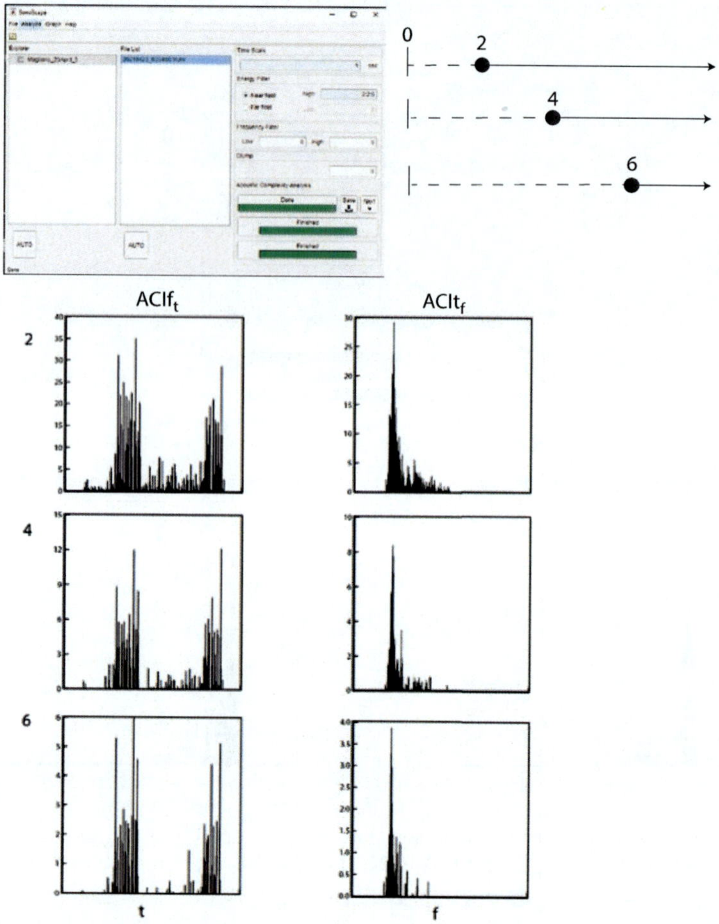

Fig. 4.11 Temporal and spectral sonic signature with the application of an energy filter = 2 to 6 with a step =2. (Magliano (44°14′ N, 10°03′ E), April 23, 2021, at 05:54 a.m., recording station #16)

4.7 Working with the Temporal Scale

In this exercise, we will guide to the calculus of ACI metrics in the simplest way by processing data of a file of 300 s. In particular, we will apply a different scale time (from 1 to 5 s) to observe the behavior of ACI metrics according to the different temporal intervals and finally to evaluate the fractal dimension of EE and of SSD (Figs. 4.20 and 4.21).

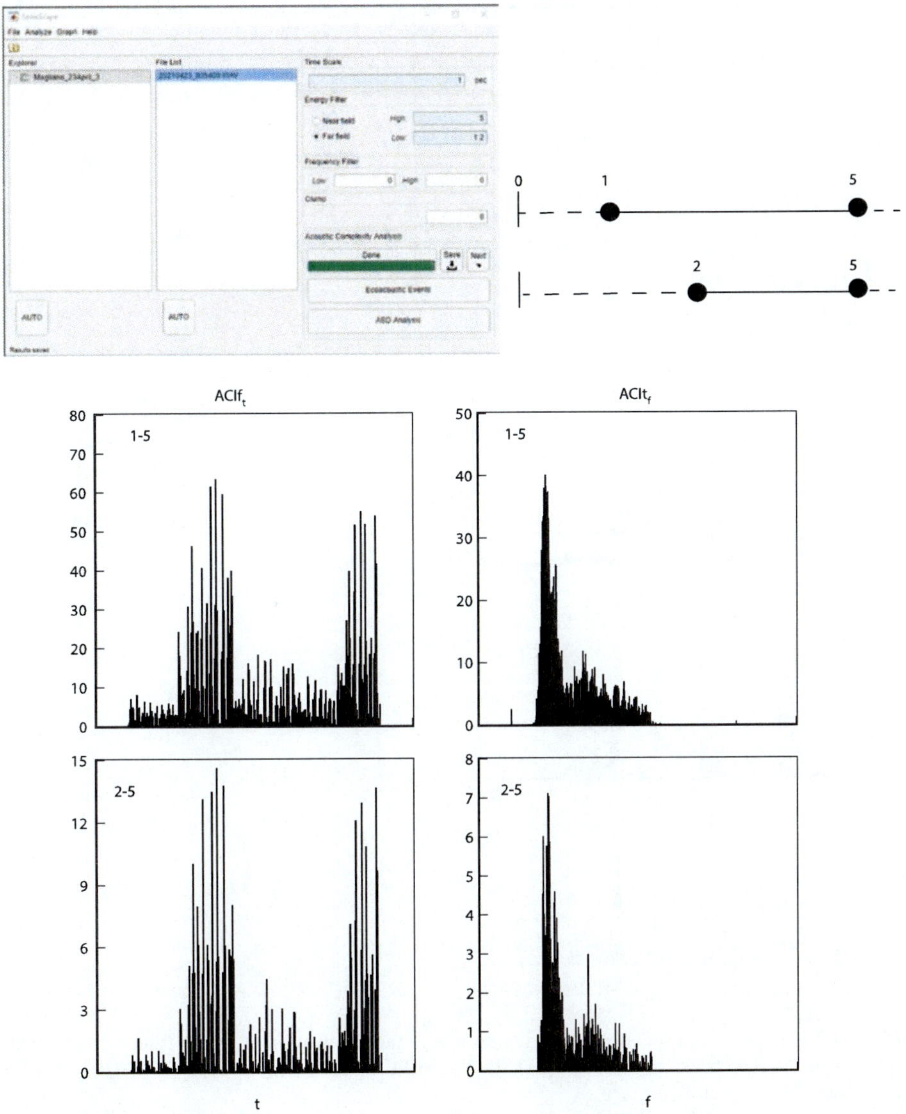

Fig. 4.12 Temporal and spectral sonic signature with the application of a near filter = 5 and a far filter =1:2. (Magliano (44°14′ N, 10°03′ E), April 23, 2021, at 05:54 a.m., recording station #16)

4.8 Working with the Fractal Dimension of the Ecoacoustic Events at Dawn

In this example, we have utilized the data from 11 (red dots) of the 11 stations deployed at a distance of 200 m between each pair (Fig. 4.22) at Magliano locality (44°14′ N, 10°03′ E). Sound was recorded at dawn for 5 min and sampled at

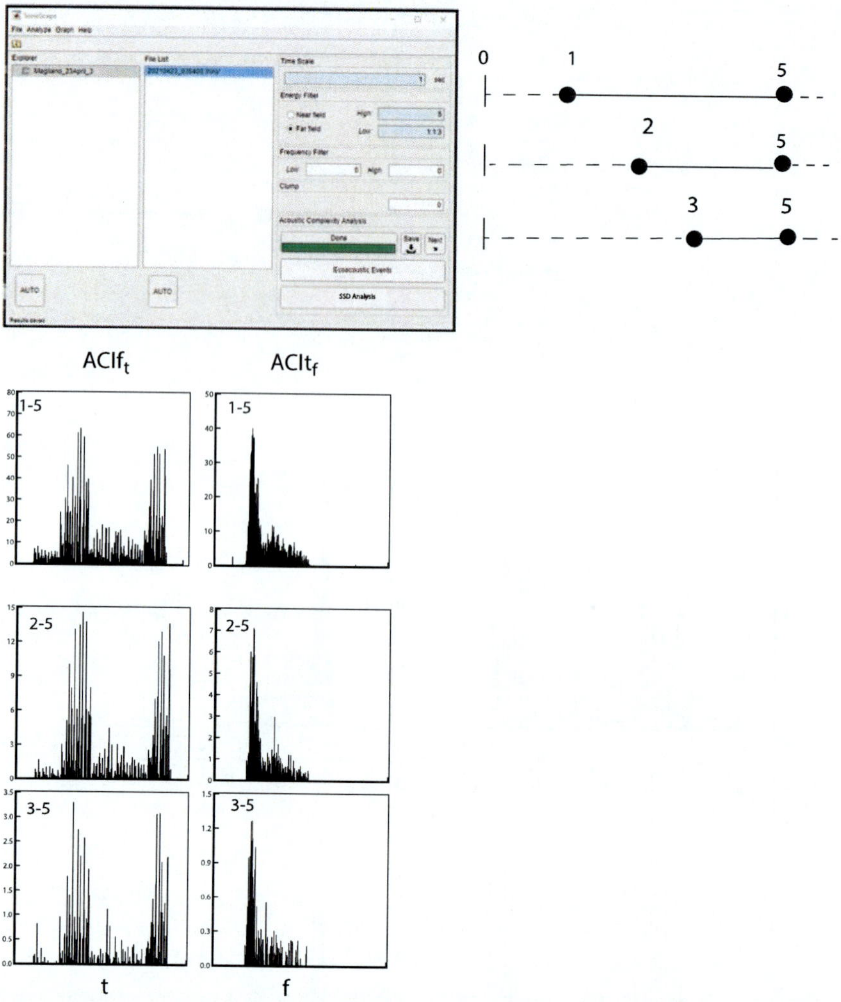

Fig. 4.13 Temporal and spectral sonic signature with the application of a near filter = 5 and a far filter from 1 to 3 with a step of 1 = (1:1:3) (Magliano (44°14′ N, 10°03′ E), April 23, 2021, at 05:54 a.m., recording station #16)

32 kHz. The analysis was performed by setting energy filter at 0.1, and the fractal dimension was calculated on a range of 10 temporal scales (from 1 to 10 s).

The fractal dimension of the Ecoacoustic Events was compared with the value of $ACIf_t$. From the analysis, it shows different behaviors between $ACIf_t$ characterized by strong differences among stations, and the fractal dimension reflects a small variability assigning at the dawn chorus which demonstrates a homogeneous pattern across the landscape despite different densities of acoustic signals.

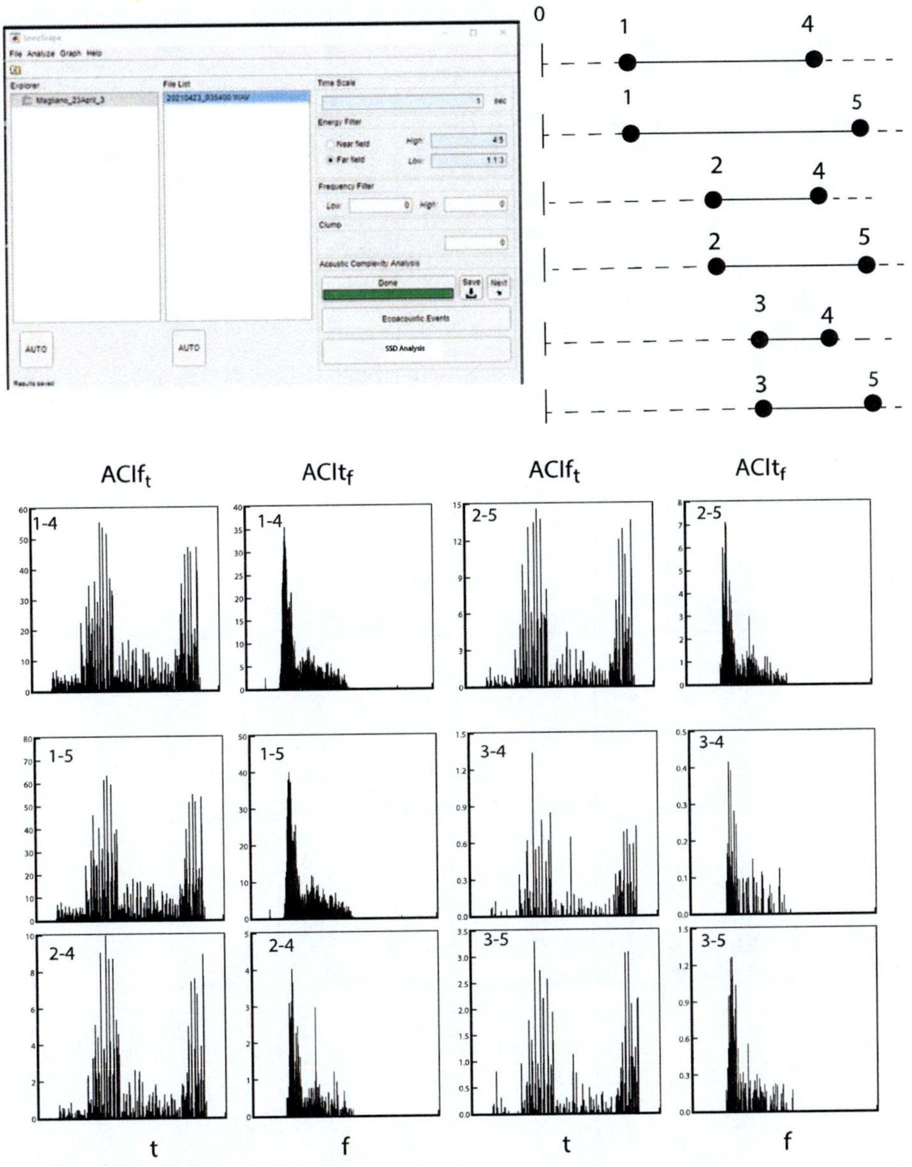

Fig. 4.14 Temporal and spectral sonic signature with the application of a near filter from 4 to 5 and a far filter from 1 to 3 with a step of 1 = (1:1:3) (Magliano (44°14′ N, 10°03′ E), April 23, 2021, at 05:54 a.m., recording station #16)

4.9 Graphing Ecoacoustics Events

When we process large files, the numerosity of Ecoacoustic codes (from 0 to 999) demands a synthetic graphical representation. To increase the manageability of such results, we have introduced a ternary plot representation. This diagram is a

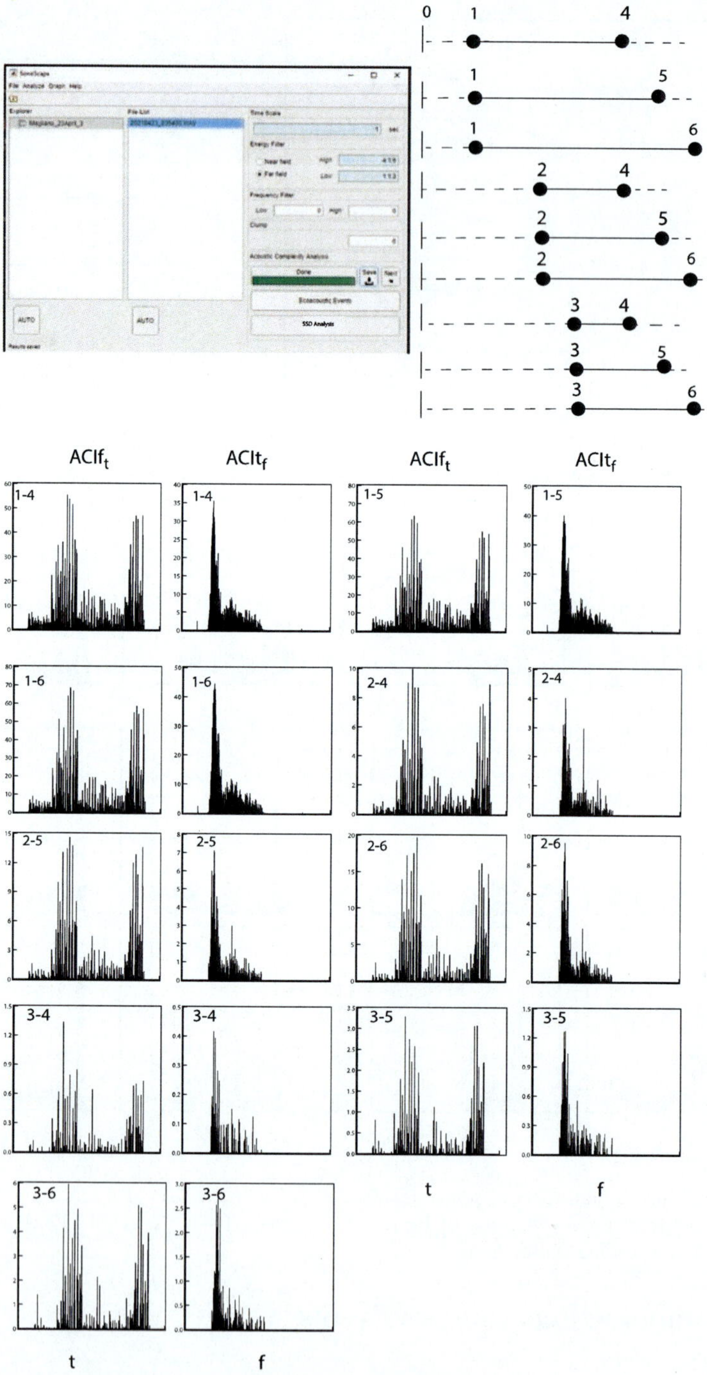

Fig. 4.15 Temporal and spectral sonic signature with the application of a near filter from 4 to 6 with step of 1 and a far filter from 1 to 3 with a step of 1 = (1:1:3) (Magliano (44°14′ N, 10°03′ E), April 23, 2021, at 05:54 a.m., recording station #16)

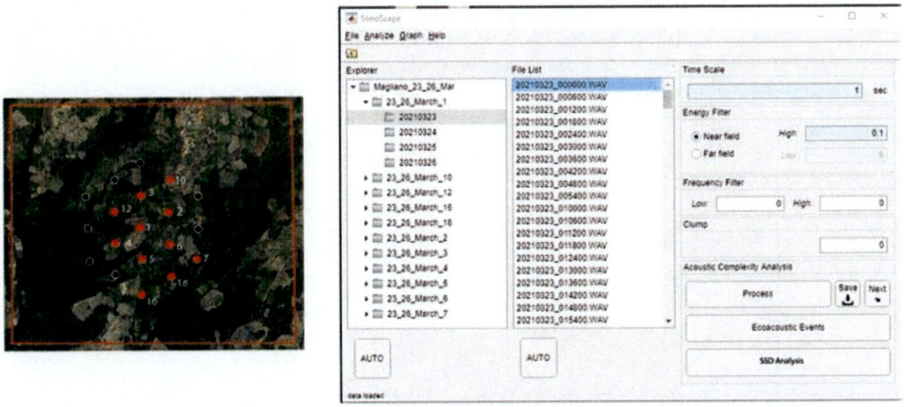

Fig. 4.16 Location of the recording stations at Magliano (44°14′N, 10°03′ E) study site and the screen shot of the main window utilized to process the nested folders, during 4 days in 11 stations labeled 1, 2, 3, 4, 5, 6, 7, 10, 11, 12, 16, and 18

barycentric plot on three variables which sum to a constant. The three variables are $ACIf_t$, $ACIf_t$ evenness, and $ACIt_f$ evenness represented by a three-digit code.

The ternary plot allows to observe the behavior of Ecoacoustic codes inside a temporal space that may be set according to the goal.

The ternary plot allows to observe the distribution of $ACIf_t$, $ACIf_t$ evenness, and $ACIt_f$ evenness in a barycentric space. This requires the transformation of each ecoacoustic code component into a relative importance value. For instance, the code 191, which corresponds to 1 = $ACIf_t$, 9 = $ACIf_t$ evenness, and 1 = $ACIt_f$ evenness, is transformed into 0.09-0.82-0.09. The code 999 into 0.33-0.33-0.33, and the code 845 into 0.47-0.24.-0.29. Finally, the relative importance of each code component is plotted in a barycentric space.

This representation has a limit around the perfect barycentric position that can be occupied by different ecoacoustic codes (111, 222, 333, 444, 555, 666, 777, 888, or 999) (Fig. 4.23a).

We have prepared six examples:

Example #1

In this example, we have utilized a sonic file from the station #16 of Magliano (44°14′ N, 10°03′ E) study site on March 23, 2021, at 4:42 a.m.

Selecting three levels of energy filter 0.02, 0.05, and 0.1 in Fig. 4.24, the distribution in the ternary plot space of ecoacoustic codes is summarized. The graphical representation shows the importance of the energy filter in the coding process of Ecoacoustic Events. Low energy filter is central to understanding the dynamics of sonoscapes. In this case, when the energy filter is very low, $ACIf_t$, $ACIf_t$ evenness, and $ACIt_f$ evenness are close to an even distribution. This depends on the fact that the majority of "random signals" enter in the computation of $ACIf_t$ masking patterns. If the filter is more severe (0.05), a differentiation of Ecoacoustic Events is

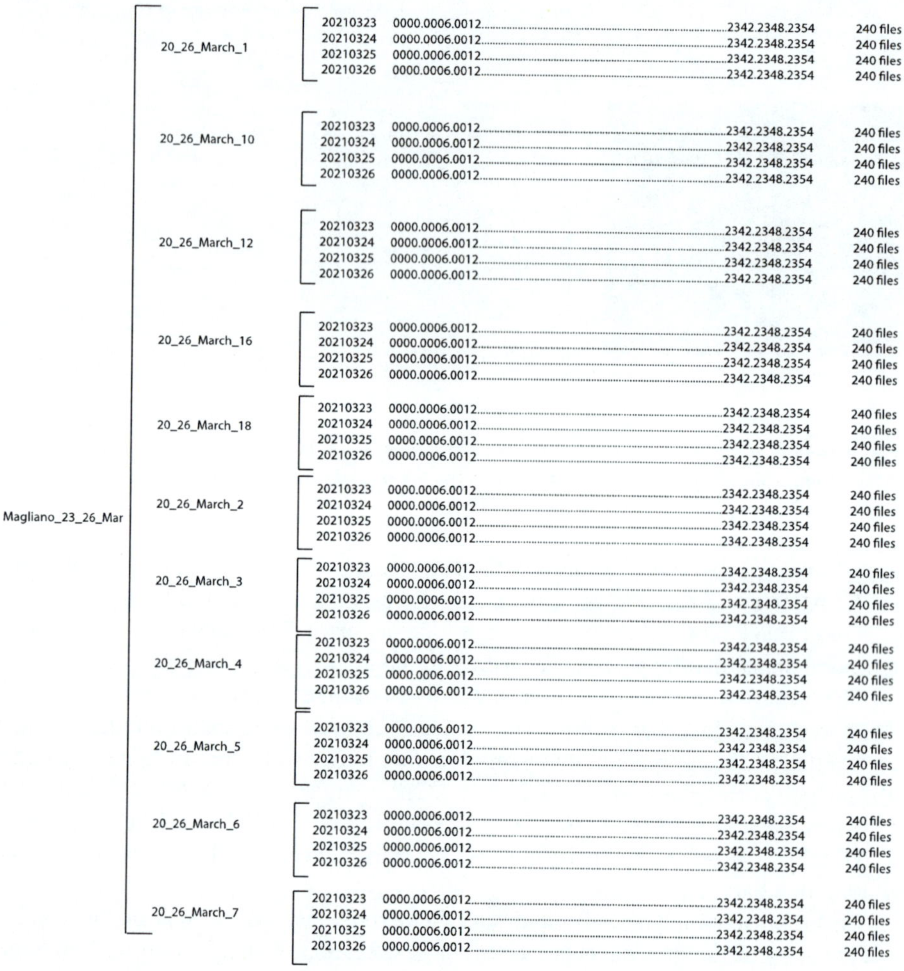

Fig. 4.17 Folder hierarchy utilized to process automatically 4 days of data from 11 sampling stations

evident. With an energy filter of 0.1, the distribution of Ecoacoustic Events categories changes with a major dispersion.

Example #2

In Fig. 4.25 are represented the Ecoacoustic codes and the sonic signature (Fig. 4.26) of 10 typologies of wind sorted according to the ACIf$_t$ value from Alaska (Courtesy of Tim Mullet), processed with an energy filter of 0.01. It appears evident that also soundscapes considered "noisy" or "without signals" have inherent relevant information that can be distinguished utilizing EE approach and graphical ternary plots.

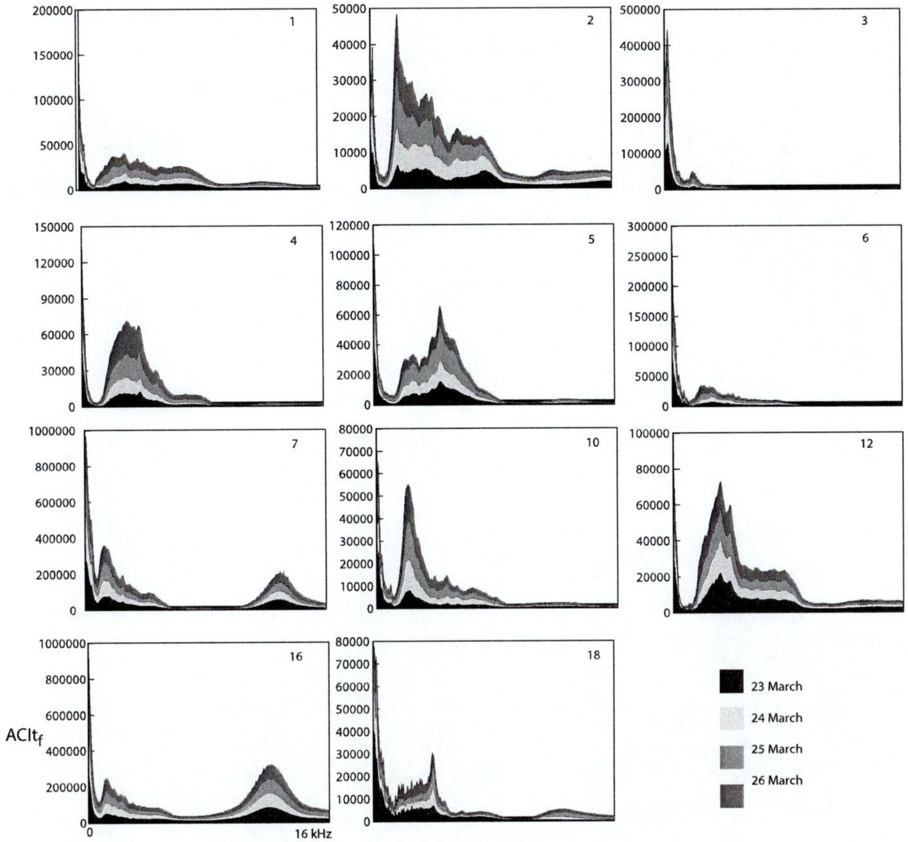

Fig. 4.18 Representation of the Sonic signature $ACIt_f$ of 11 stations (1, 2, 3, 4, 5, 6, 7, 10, 12, 16, and 18) in Magliano locality (44°14′ N, 10°03′ E) obtained by recording for 300 s 240 time a day along 4 days. The profiles are stacked together day over day in order for better comparing the patterns. And because of this, the absolute value of the *y* axis becomes less meaningful

Example #3

In Fig. 4.27 are represented the Ecoacoustic codes and the sonic signature (Fig. 4.28) of 10 typologies of rain from Alaska (Courtesy of Tim Mullet). It appears evident that also soundscapes considered "noisy" or "without signals" have relevant information. With the increase in rain, the ternary codes tend to move at the center of the triangle indicating a barycentric position for $ACIf_t$, $ACIf_t$ evenness, and $ACIt_f$ evenness.

Example #4

In Fig. 4.29 are represented the Ecoacoustic codes and the sonic signature of 10 typologies of "silent sonoscape" from Alaska (Courtesy of Tim Mullet). It also appears evident that sonoscapes considered "noisy" or (in opposition) "without

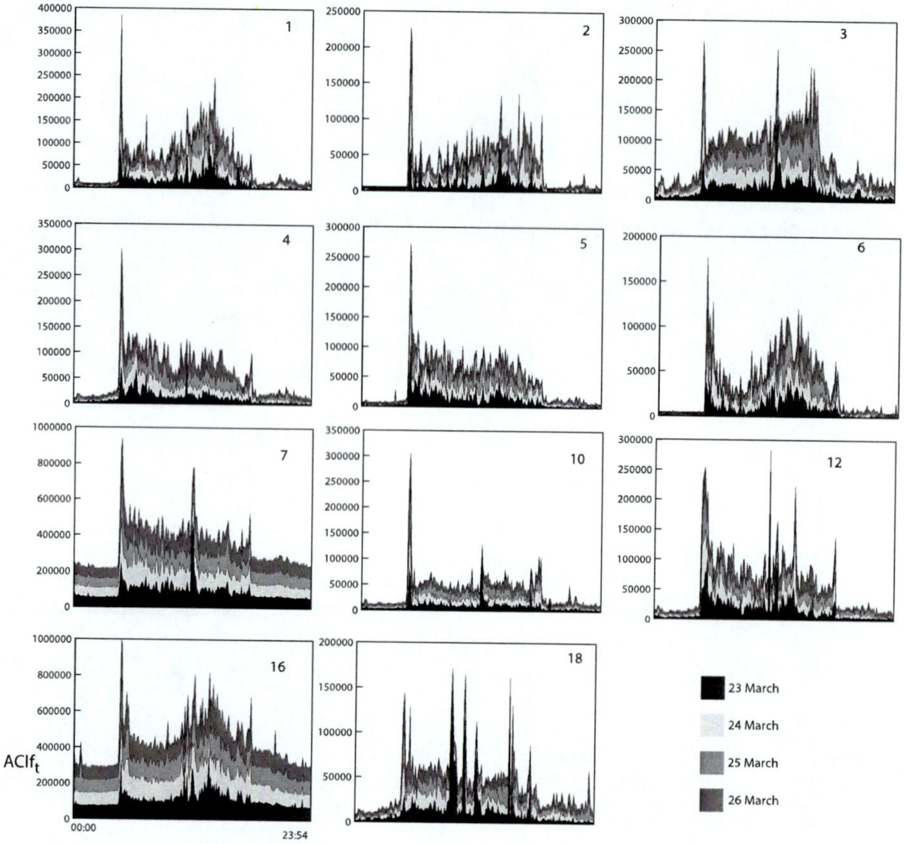

Fig. 4.19 Representation of the temporal sonic signature ACIf$_t$ of 11 stations (1, 2, 3, 4, 5, 6, 7, 10, 12, 16, and 18) in Magliano study area (44°14′ N, 10°03′ E) obtained by recording for 300 s 240 time a day along 4 days. The profiles are stacked together day over day in order for better comparing the patterns. And because of this, the absolute value of the *y* axis becomes less meaningful

signals" have some information that can be plotted, although in the cases illustrated in Fig. 4.29, the differences emerging from the ternary plots seem really negligible.

Example #5

In this example, we have utilized 11 five-min sonic files from Magliano study area (44°14′ N, 10°03′ E) collected on May 1, 2021, at 3:42 GMT a.m. during the maximum of dawn chorus (Fig. 4.30). The energy filter has been fixed at 0.1. The ternary plot should highlight some minor differences in frequency and time distribution according to the different stations. The relative homogeneity of Ecoacoustic codes reflects a growing homogeneity across this rural landscape due to a diffuse land abandonment and an expansion of woodland cover.

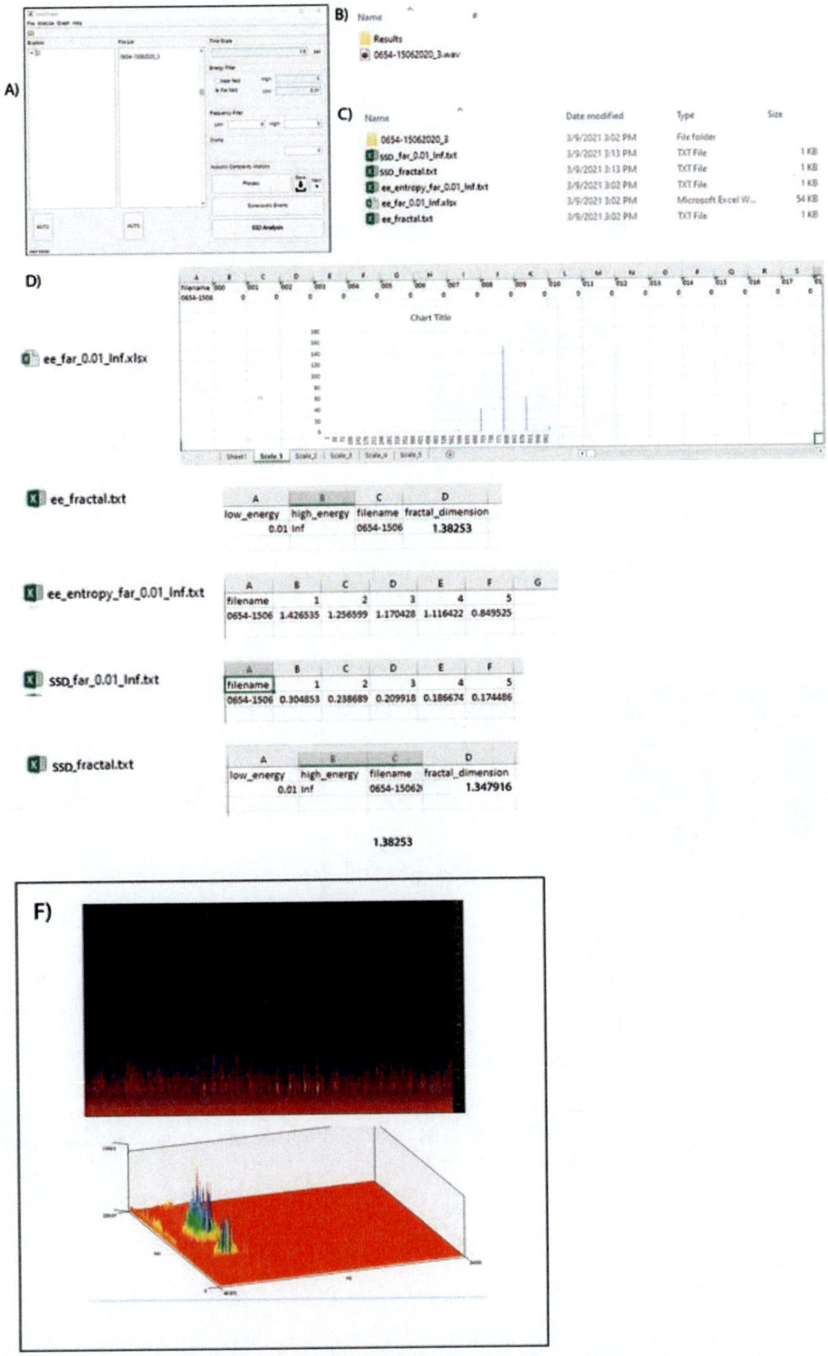

Fig. 4.20 In (**A**) is represented the window with the setting parameters for the file 0654-15062020_3.wav (hh = 06, mm = 54, day = 15, month = 06, year = 2020): time scale from 1 to 5 s (1, 2, 3, 4, 5), the energy filter of 0.01. In (**B**) the content of the folder. In (**C**) some files of the Result (with the exclusion of 0654-15062020_3 subfolder) (**D**) The details of each file (ee_far, ee_fractal, ee_entropy, ssd, ssd_fractal). (**F**) The spectrogram and its 3D representation

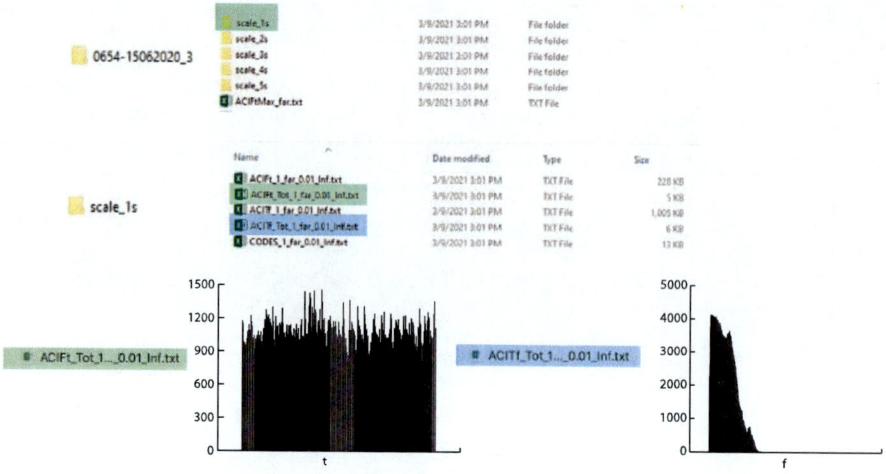

Fig. 4.21 Continuing from Fig. 4.20, the subfolder 0654-15062020_3 is composed of a number of folders equal to the scaling (in this case 5). In each scaling folder are deposited two matrices: $ACIf_t$ (46 values × 3 00 s) and $ACIt_f$ (300 s × 512 spectral lines), and their respective summaries are represented in the two graphics

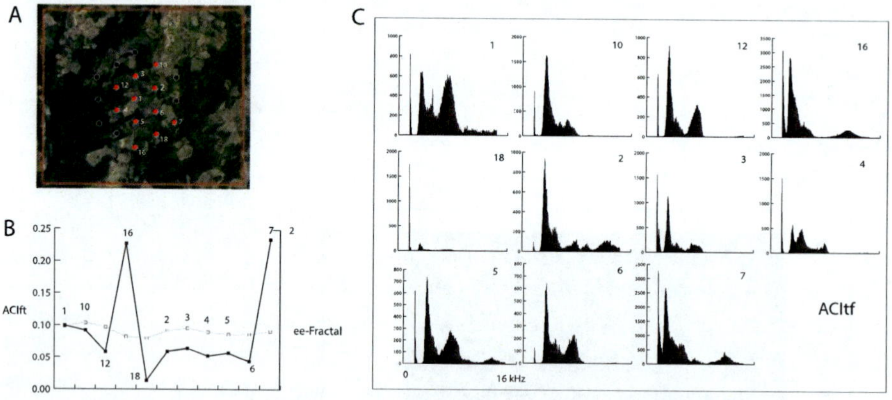

Fig. 4.22 (**A**) Spatial arrangement of 11 recording stations (red dots) on a net of 19, across a rural landscape in Magliano (44°14′ N, 10°03′ E). (**B**) $ACIf_t$ and Ecoacoustic event fractal dimension (ee-Fractal). (**C**) Sonic signatures ($ACIt_f$) in the 11 recording stations

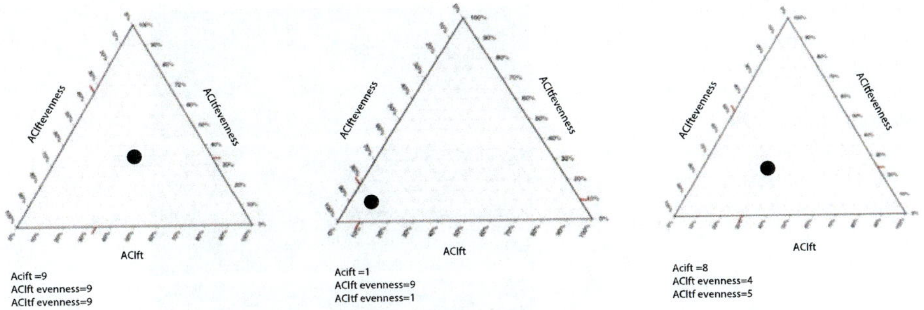

Fig. 4.23 Distribution of different codes (999, 191, 485) in a ternary plot

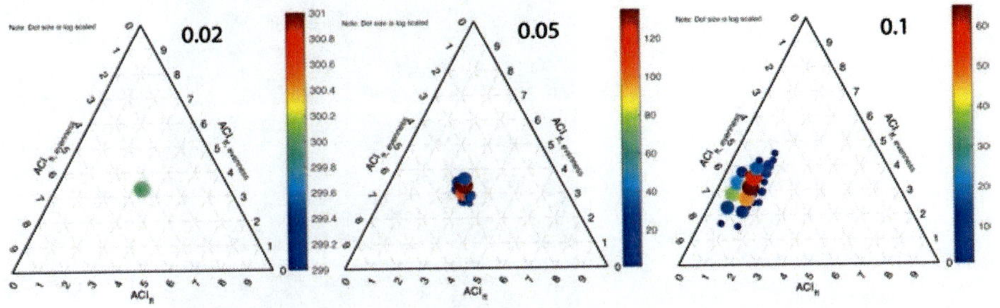

Fig. 4.24 Graphical representation of EE along three energy filters (0.02, 0.05, and 0.1)

Example #6

In this example, we have utilized 11 five-min sonic files from Pranda lake study area (44°18′ N, 10°14′ E) collected on May 2, 2021, at 3:36 GMT during the maximum of dawn chorus (Fig. 4.31). The energy filter has been fixed at 0.1. The ternary plot shows some relevant differences in frequency and time distribution according to the different stations. The presence of evident differences in EE distribution is a signal of a major environmental constraint that in turn differentiates distribution and song activity of the soniferous community. Stations #11 and #13 present a barycentric position of ecoacoustic codes due to the only presence of geophonies.

In conclusion, ternary plot graphic greatly improves the capacity of analysis of the Ecoacoustic events and opens the way to a more precise analysis saving time and gaining in efficiency.

Modulating the temporal resolution that modifies the number of temporal steps analyzed is possible to reduce the number of EE concentrating the visibility of the results as in the example of Fig. 4.32, where the temporal scale of the data previously elaborated in Fig. 4.31 has been changed in periods of 60 s.

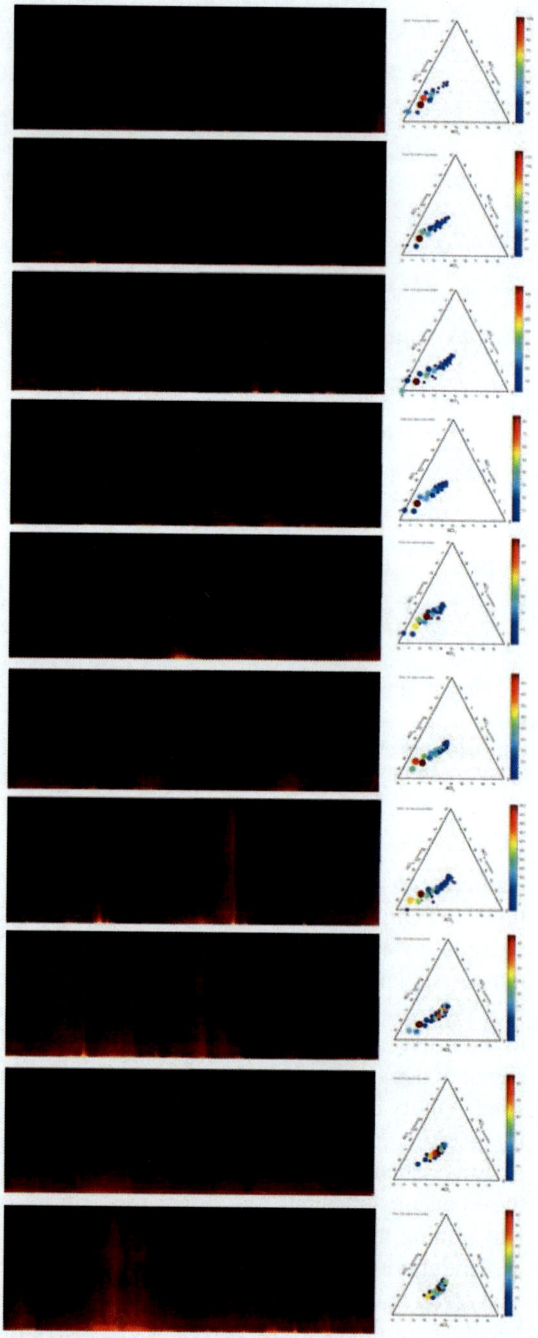

Fig. 4.25 Ten typologies of wind recorded in an Alaska forest and sorted for $ACIf_t$ increasing from top to bottom: spectrograms and ternary plot (data are from courtesy of Tim Mullet)

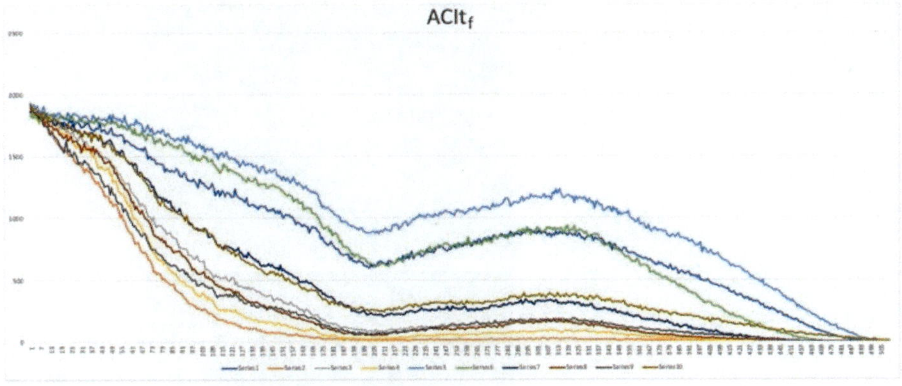

Fig. 4.26 Distribution of the sonic signatures of 10 conditions of wind in Alaska (Courtesy of Tim Mullet)

The differences between the different localities are based on 300/60 = 5 EE code for each files, compared with 300 codes adopting a scale of 1 s.

The ternary plot helps to better understand the evolution of the sonoscapes and allows to make precise evaluations especially for situations poor in biophonic signals. In particular, in the investigation on geophonies, it seems realistic to categorize such sounds which, until now, have been considered as only noise, and as such, it is always erased from a spectrogram.

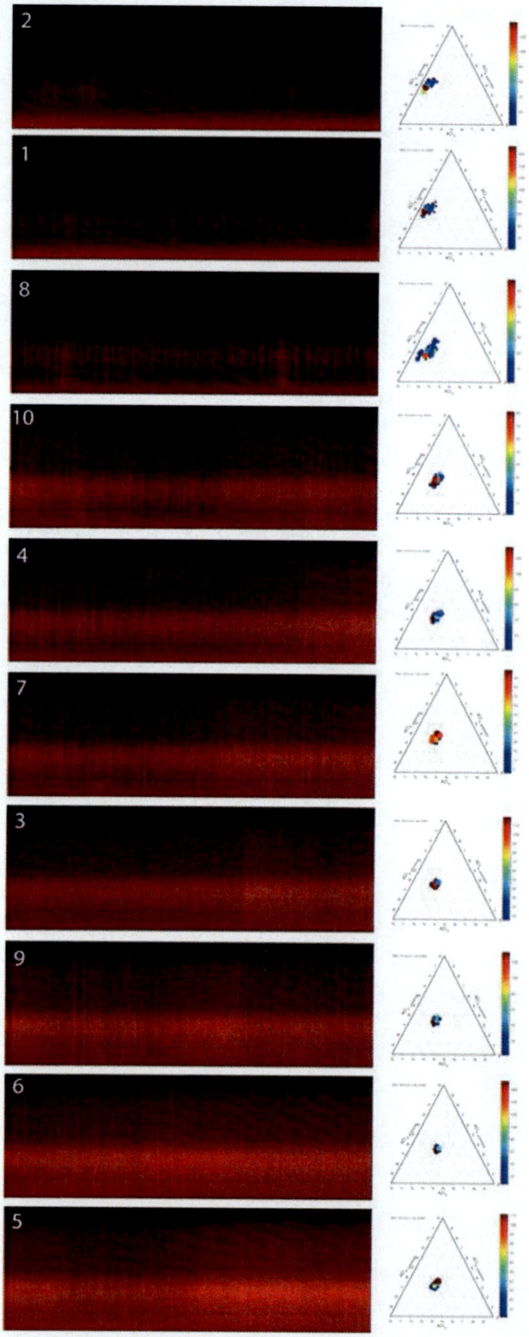

Fig. 4.27 Ten typologies of rain from an Alaska forest and sorted for $ACIf_t$ increasing from top to bottom. The numbers are the file identifier (Courtesy of Tim Mullet)

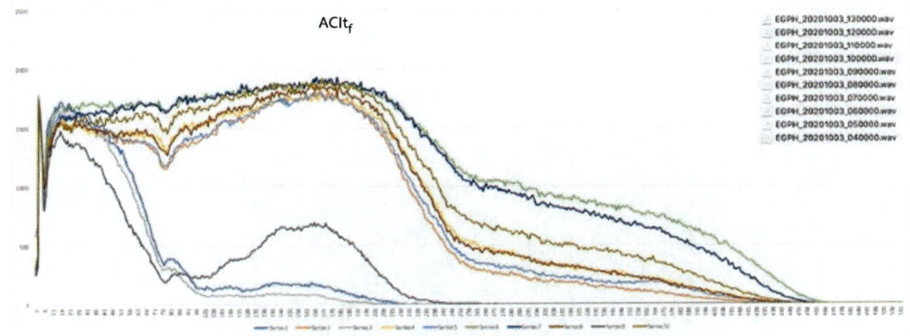

Fig. 4.28 Sonic signature (ACIt$_f$) of 10 typologies of Alaska rain (Courtesy of Tim Mullet)

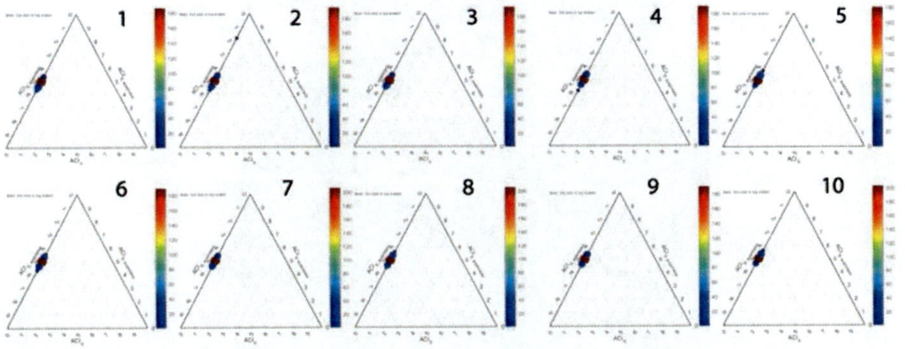

Fig. 4.29 Alaska Silent. The first two spectral lines have been excluded to improve the visibility of Ecoacoustic codes in the ternary plot. The sonic signature of individual files overlaps due to a strong similarity between the different files that contain very few and faint sonic signals

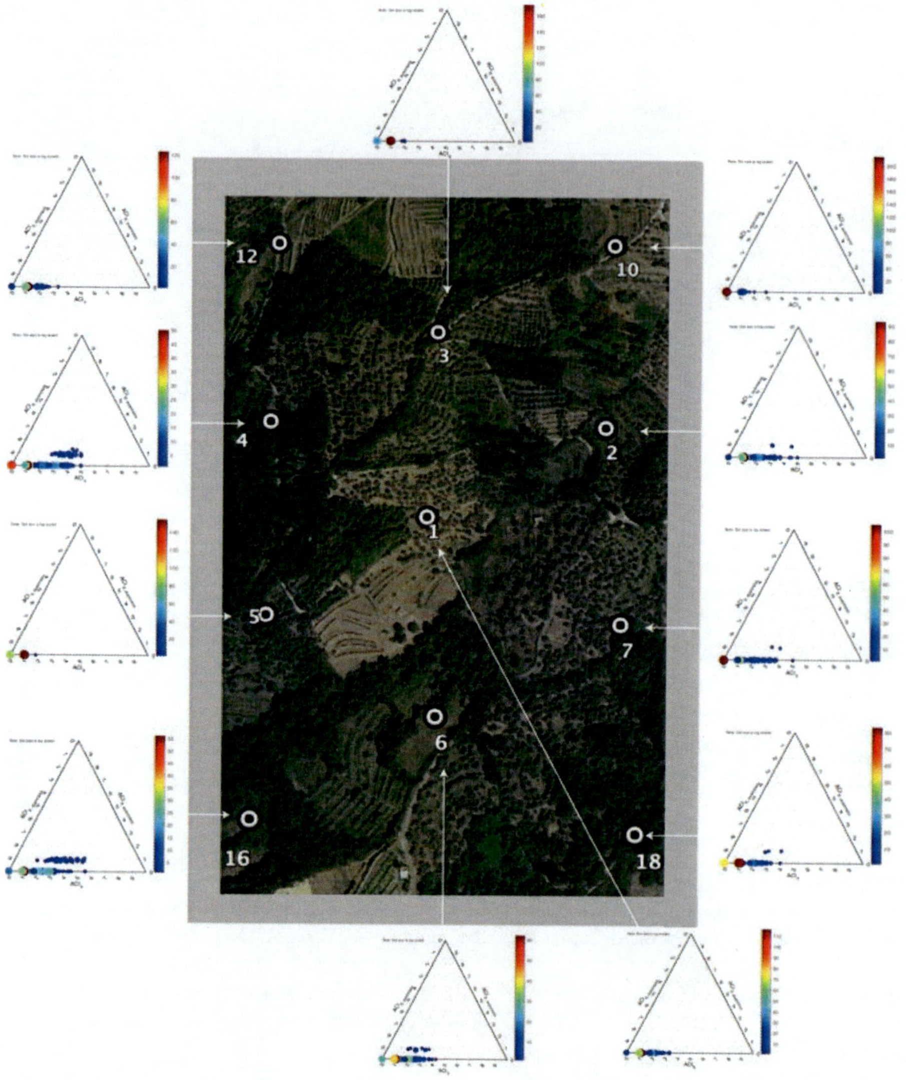

Fig. 4.30 Distribution of Ecoacoustic codes in 11 recording stations at Magliano study area (44°14′ N, 10°03′ E) on May 1, 2021

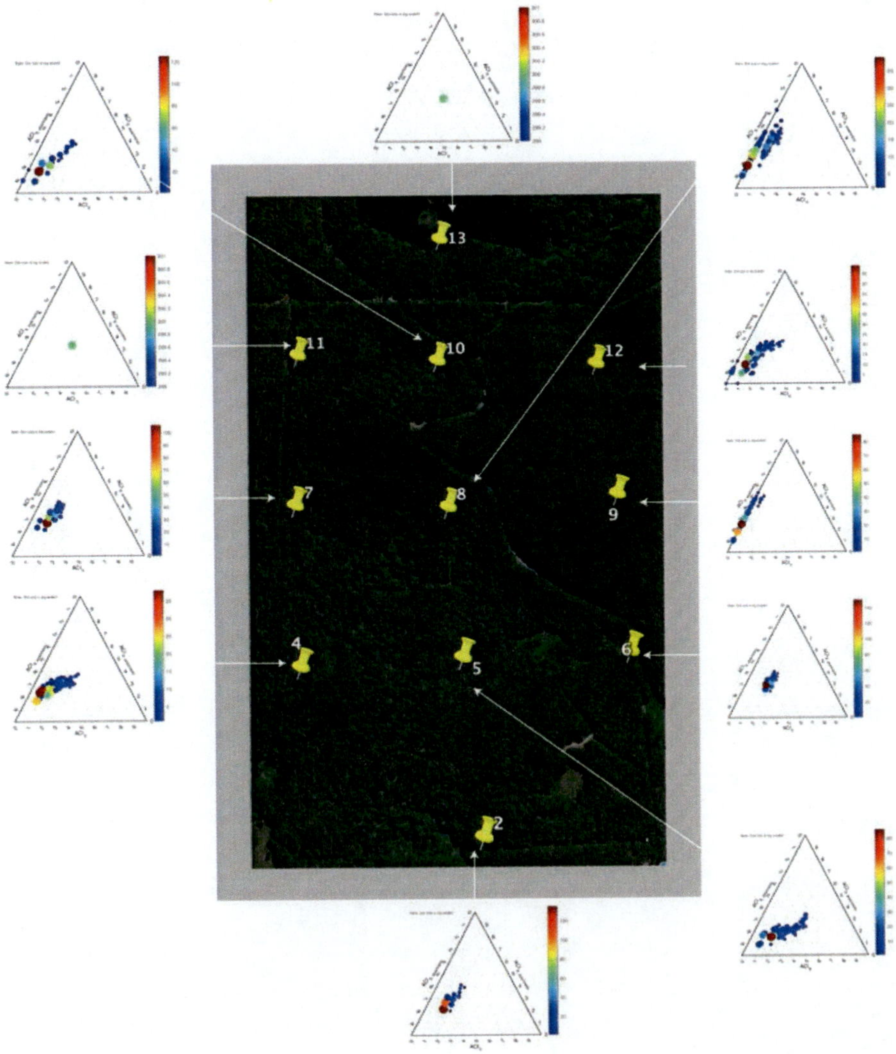

Fig. 4.31 Distribution of ecoacoustic codes in 11 recording stations at Pranda lake study area on May 2, 2021 (44°18′ N, 10°14′ E)

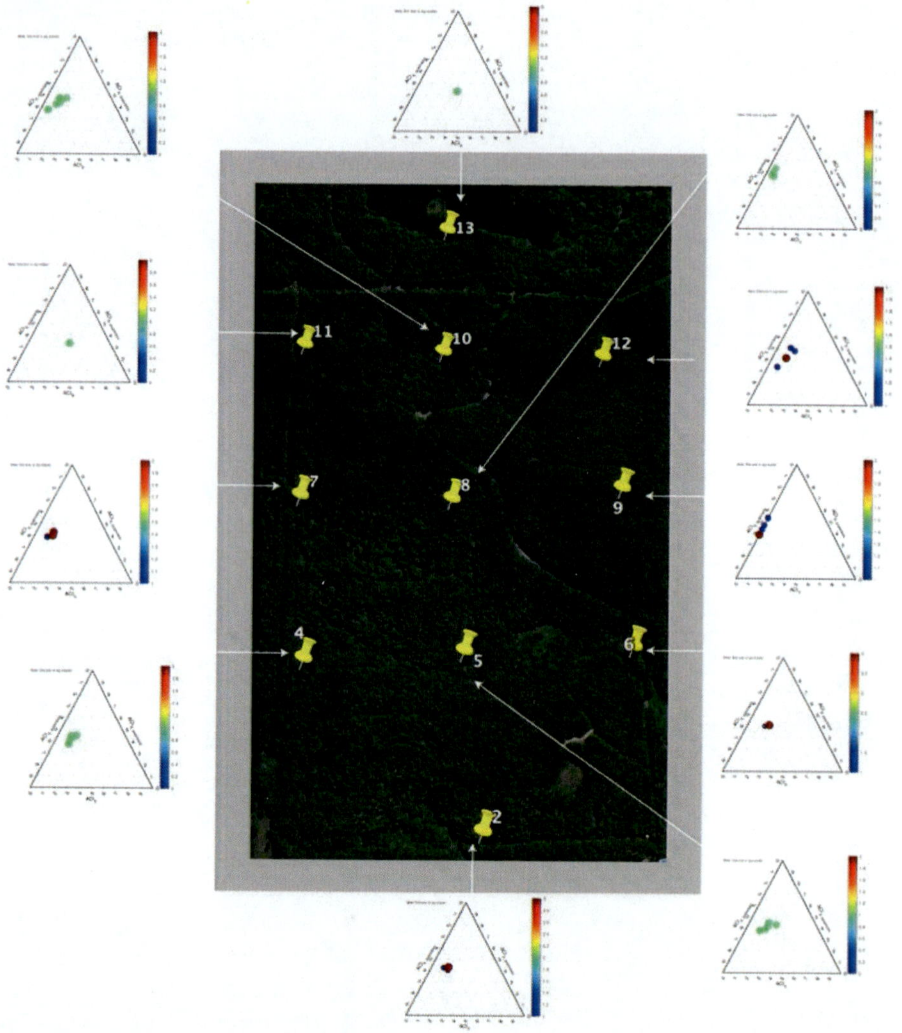

Fig. 4.32 Distribution of Ecoacoustic code in 11 recording stations at Pranda lake study area (44°18′ N, 10°14′ E) on May 2, 2021, according to a temporal scale of 60″

Index

A. Farina, P. Li, *Methods in Ecoacoustics*, Frontiers in Ecoacoustics 1,
https://doi.org/10.1007/978-3-030-82177-7